Engineering Iron and Stone

Other Titles of Interest

America Transformed: Engineering and Technology in the Nineteenth Century, by Dean Herrin. (ASCE Press, 2003). Displays a visual sampling of engineering and technology from the 1800s that demonstrates the scope and variety of the U.S. industrial transformation. (ISBN: 9780784405291)

History of the Modern Suspension Bridge, by Tadaki Kawada, Ph.D.; translated by Harukazu Ohashi, Ph.D.; and edited by Richard Scott, M.E.S. (ASCE Press, 2010). Traces the modern suspension bridge from its earliest appearance in Western civilization only 200 years ago to the enormous Akashi Kaikyo and Storebaelt bridges completed at the end of the twentieth century. (ISBN: 9780784410189)

Circles in the Sky, by Richard G. Weingardt, P.E. (ASCE Press, 2009). Chronicles the life of George Ferris, the civil engineer and inventor responsible for creating, designing, and building the Ferris Wheel. (ISBN: 9780784410103)

Structural Identification of Constructed Systems, edited by F. Necati Çatbas, Ph.D., P.E.; Tracy Kijewski-Correa, Ph.D.; and A. Emin Aktan, Ph.D. (ASCE Technical Report, 2013) Presents research in structural engineering that bridges the gap between models and real structures by developing more reliable estimates of the performance and vulnerability of existing structural systems. (ISBN: 9780784411971)

Engineering Iron and Stone

Understanding Structural Analysis and Design Methods of the Late 19th Century

Thomas E. Boothby, Ph.D., P.E.

Library of Congress Cataloging-in-Publication Data

Boothby, Thomas E.
Engineering iron and stone : understanding structural analysis and design methods of the late 19th century / Thomas E. Boothby, Ph.D., P.E.
pages cm
Includes index.
ISBN 978-0-7844-1383-8 (print : alk. paper)—ISBN 978-0-7844-7894-3 (ebook)—ISBN 978-0-7844-7895-0 (epub) 1. Building, Iron and steel—History—19th century. 2. Building, Stone—History—19th century. 3. Structural analysis (Engineering)—History—19th century. I. Title.
TA684.B736 2015
624.1′82109034—dc23

2014040873

Published by American Society of Civil Engineers
1801 Alexander Bell Drive
Reston, Virginia 20191
www.asce.org/pubs

Any statements expressed in these materials are those of the individual authors and do not necessarily represent the views of ASCE, which takes no responsibility for any statement made herein. No reference made in this publication to any specific method, product, process, or service constitutes or implies an endorsement, recommendation, or warranty thereof by ASCE. The materials are for general information only and do not represent a standard of ASCE, nor are they intended as a reference in purchase specifications, contracts, regulations, statutes, or any other legal document. ASCE makes no representation or warranty of any kind, whether express or implied, concerning the accuracy, completeness, suitability, or utility of any information, apparatus, product, or process discussed in this publication, and assumes no liability therefor. The information contained in these materials should not be used without first securing competent advice with respect to its suitability for any general or specific application. Anyone utilizing such information assumes all liability arising from such use, including but not limited to infringement of any patent or patents.

ASCE and American Society of Civil Engineers—Registered in U.S. Patent and Trademark Office.

Photocopies and permissions. Permission to photocopy or reproduce material from ASCE publications can be requested by sending an e-mail to permissions@asce.org or by locating a title in ASCE's Civil Engineering Database (http://cedb.asce.org) or ASCE Library (http://ascelibrary.org) and using the "Permissions" link.

Errata: Errata, if any, can be found at http://dx.doi.org/10.1061/9780784413838.

Copyright © 2015 by the American Society of Civil Engineers.
All Rights Reserved.
ISBN 978-0-7844-1383-8 (print)
ISBN 978-0-7844-7894-3 (PDF)
ISBN 978-0-7844-7895-0 (EPUB)
Manufactured in the United States of America.

22 21 20 19 18 17 16 15 1 2 3 4 5

Cover credits: (Front cover) Cabin John Bridge schematic courtesy of Special Collections, Michael Schwartz Library, Cleveland State University. Cabin John Bridge photo (2014) by David Williams.

(Back cover) Cabin John Bridge watercolor: Library of Congress, Prints & Photographs Division, Historic American Engineering Record, Reproduction No.: HAER MD,16-CABJO,1—12 (CT). Cabin John Bridge photo (August 1861): Library of Congress, Prints & Photographs Division, Historic American Engineering Record, Reproduction No.: HAER MD,16-CABJO,1—10.

This book is affectionately dedicated to
Colin Bertram Brown
1929–2013

But O for the touch of a vanish'd hand,
And the sound of a voice that is still!
Alfred, Lord Tennyson

Contents

Preface

This book stems from a career-long interest in understanding how structural engineers worked in the past. Although we admire the great works of Roman engineering and the medieval cathedrals of Europe, we tend to think that modern engineering is somehow superior to the engineering that produced these structures. The premise of this book is that, for all its evident differences, modern engineering cannot claim superiority to the engineering of any period in the history of civilization. That contemporary engineering is based on a different mindset and a different set of values from the work of any of these other periods is evident. But the works that appeared in the engineering of other periods are not reproducible by contemporary methodology: each age defines its own artifacts and its own ways of producing these artifacts.

The late nineteenth century is a particularly significant time for understanding contemporary engineering: Although nineteenth-century engineering is different from modern engineering in the sense described, this period is closely related to the present time. Although Roman and medieval engineering are defined primarily by experience-based procedures, they are somewhat informed by emerging ideas from speculative science. By the nineteenth century, however, ideas of science were sufficiently advanced, and ideas about the role of science in society, such as positivism, were sufficiently widespread that engineers began to think of themselves as scientists of a sort and began to think that they were responsible for applying scientific procedures to constructed works.

A particularly interesting feature that emerged from the study of nineteenth-century engineering methods was the efficiency and accuracy of some of the procedures employed, as compared with the way we accomplish these tasks in the present age. Particularly in truss design, both analytical and graphical, most of the procedures employed in the nineteenth century appear to be more efficient than those that we teach to students in contemporary engineering programs. The reliance on graphical methods, especially for trusses and arches, is particularly revealing of the late nineteenth-century mindset and does influence the actual form of the structures.

In preparing this book I tried to focus on ordinary procedures used to design and construct ordinary works without placing emphasis on the exceptional engineering works that mark this period. Thus, although the reader can find references to the design of major works, most of the discussions in this book describe smaller works and the significant body of engineering design that went into their construction.

Acknowledgments

I have been assisted greatly in many ways by many people in the preparation of this book. I have received particular assistance from several libraries that I would like to acknowledge. Daniel Lewis at the Huntington Library, San Marino, CA, has been particularly helpful, as have all the staff at the Avery Library at Columbia University, Ilhan Citak at the Linderman Library at Lehigh University, and the Special Collections staff at the Penn State University Libraries. I would like to acknowledge the assistance I have received from the staff at ASCE Press, particularly from Betsy Kulamer, Donna Dickert, and Sharada Gilkey. I note the editorial assistance I have received from Mary Byers and from my brother, Daniel Boothby. I am also grateful for the support and assistance I have received from my colleagues, notably Jeffrey Laman, Louis Geschwindner Jr., Harry West, and Theodore Galambos. I am very grateful to Brice Ohl and Oluwatobi Jewoola, undergraduate students at Penn State University, for the preparation of the illustrations found throughout the book. I have received continual help and encouragement from my friends at the Engineering Copy Center, Penn State University. Finally, I gratefully acknowledge the patience, comfort, and help of my wife, Anne Trout, over the four years during which this book was developed.

Introduction

This book concerns the methods used for structural engineering design in the late nineteenth century. Even as the opportunities for business, industry, and transportation were expanding during this time, the methods of the civil engineering and the structural engineering professions were also expanding, in part to meet the demands of the expansion of industry. The intent of the present book is to capture, through investigation of writings, archival evidence, and examination of built works, the methods of structural design of bridges and buildings in the period from 1870 through 1900, roughly, the period known to historians as the Gilded Age (1865–1893). The value of this exercise is three-fold. First, understanding the intent of the designer is the key to a successful rehabilitation, whether architectural or structural. Second, the preservation of design methods for historic structures is at least as important as the preservation of the structures themselves. Third, many of the methods used in structural design in the late 1800s are valuable in their own right—quick, computationally efficient, understanding of the behavior of the structure, and often giving special insight into the actual performance of the structure.

In undertaking the historic preservation of structures from the late nineteenth century, understanding design intent is important—the way that a bridge or building was designed and the way that the elements of the structure were intended to function. Too often in historic preservation projects, we overlook the designer's conception of the structure and impose a modern outlook on the structure, with the result that significant historic fabric is removed unnecessarily. One of the most widespread misunderstandings concerning historic structures is the idea that the older structures were designed for lighter loading than modern-day structures. In fact, road bridges were designed for deck loads of up to 100 lbs/ft^2 (see, for instance, Waddell 1894); the 1,000 lb/ft on a 10-ft lane dictated by this loading is well above the lane loading requirements of AASHTO HS-20 (AASHTO 2013). Extraordinary vehicles, such as freight drays and road rollers, imposed very heavy loads on bridges. A passage of a steamroller is illustrated in the photo of the circa 1890 opening of the St. Mary's Street Bridge in San Antonio, TX (Figure I-1). Equally important is understanding in exactly what way nineteenth-century bridge design may have differed from modern design. Although most bridge decks do meet the AASHTO uniformly distributed lane load requirement, few nineteenth-century bridge designers imposed limits on the concentrated loads that the bridge could resist. A distributed load of 100 lbs/ft^2 placed to create maximum force in each member was usually the only loading requirement. As a result, focusing attention on the floor system of a bridge under rehabilitation is more important than on the main load-carrying system, such as truss, girder, or suspension cable.

Building floor loads used in the nineteenth century were similar to those used today. However, the approach to wind loads on buildings was very different. Because much heavier roof structures were present, uplift of the roof structure generally was not considered to be a design issue, although the possibility of wind loads causing a force reversal in a web member of a truss was considered by applying wind pressure to the windward side of a roof and by removing all load from the lee side.

Although the primary intent of this book is simply to present the methods of late nineteenth-century structural design and to recognize the inherent truth, simplicity, and value

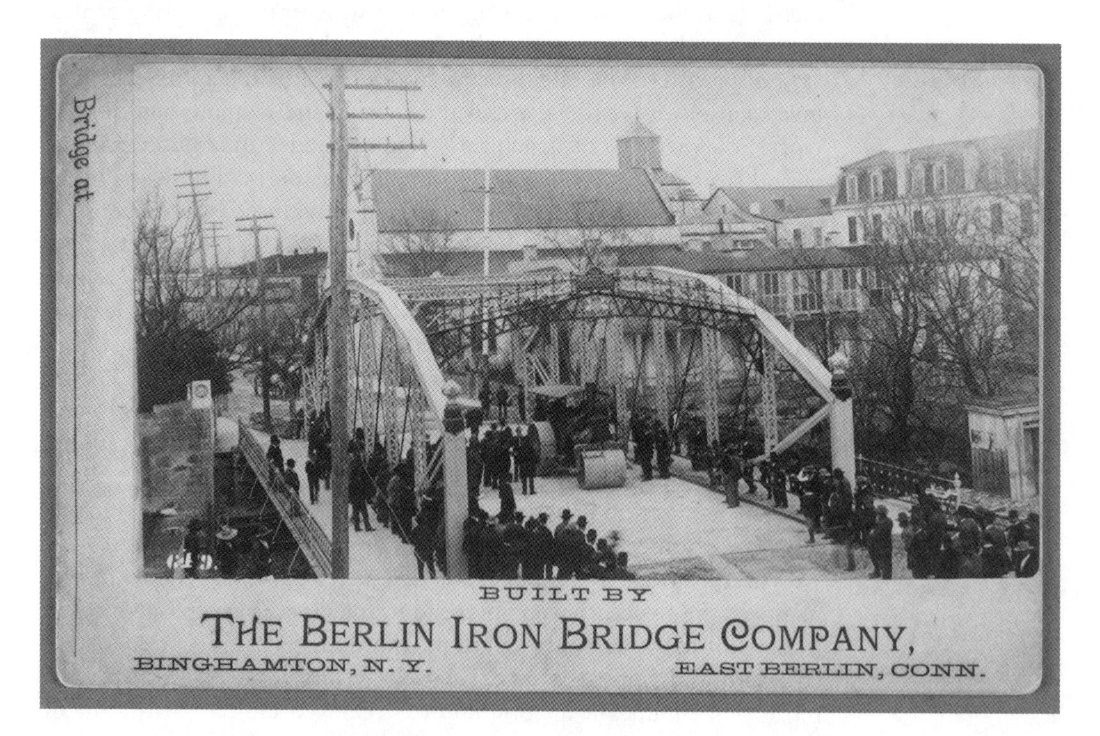

Figure I-1. Opening of St. Mary's Street Bridge, San Antonio, TX, circa 1890.
Source: Reproduced by permission of the Huntington Library, San Marino, CA.

of these methods, greater sympathy and understanding for the methods by which a structure was designed may follow directly from the review of these methods. For all the merit of contemporary engineering analysis, it is worth considering from the outset that the designers of the original structure probably knew what they were doing. In evaluating the notion that many shorter span masonry bridges were designed empirically, understanding the success of this design method for structures of this type is important. Some of the most admired and most enduring masonry structures in the world also were designed empirically, whereas, conversely, contemporary structural analysis is not always able to explain the behavior of these structures. A frequent response of contemporary engineers in rehabilitation projects involving masonry bridges is to find the structure deficient by some form of modern structural analysis or to declare it "unrateable" and in need of reinforcement by saddling the arch or installing internal anchors. This response may be appropriate in a few cases, but it needs to result from a positive determination of why the structure is deficient, including the contradiction of the original designers' findings that this was an appropriate design, for instance, clear evidence of scour or formation of hinges in the arch ring. To say that the bridge was designed for horses and buggies is incorrect; bridge decks in urban settings usually were designed for loads of 100 lbs/ft^2, a load appropriate to the heavy vehicles that were in use at the time.

Similar arguments apply to building structures. An examination of contemporary documents reveals that the live loads in widespread use were greater than the loads used in design in contemporary codes and that the safety factors generally were greater. The underlying assumption of a rehabilitation effort could be that the original designers had it right.

Although significant recent attention has been directed toward the preservation of bridges and buildings, the ideas that are reflected in the design of a historic structure also merit preservation. For the reasons described herein, it is important that we retain the ability to understand a structure from the same viewpoint as a nineteenth-century engineer. The methods presented in this book have intrinsic value, that is, they are interesting on their own account. The methods also have comparative value: comparing the methods presented here with contemporary methods is a useful exercise. As an example, consider the Rankine-Gordon formula for column capacity (Chapter 9). This formula has a firm basis in reason, calculating the residual axial force capacity for an eccentrically loaded column. As such, it considers eccentricities without introducing the idealization of a perfectly straight, perfectly concentrically loaded column and the three curves (yield, Euler Buckling Theory, and interpolation) necessary to draw a complete column curve according to either the AISC (2011) specification for steel or the *National Design Specification for Wood Construction* (American Wood Council 2006). The methods presented in this book also have pedagogical value as an accompaniment to the current building codes and standards: it is useful to provide students with alternative means of achieving the same ultimate objective, which is to build worthy structures. This book is intended as an initial step toward the preservation of these ideas, in addition to preserving the structures themselves.

Finally, the methods outlined in this book may, in some cases, be superior to the methods used in contemporary practice. The rapidity of computation and the intimate relation between the structure and its analysis present in early methods of analysis have been lost by the numerically intensive analytical methods employed in the present. In the graphical analysis of a load-carrying structure, for instance, the forces acting on a structure, the bending moments, and a suitable shape for the structure can be inferred from a single diagram. The flow of forces through a truss under variable loading can be immediately understood using some of the analytical methods for trusses that the book will explore. Some of the historic computation methods also depend on an ability to visualize the transmission of forces through a structure that is not evident in the application of computer-based methods of analysis. In particular, graphic analysis is practically a concurrent method of analysis and design in which a diagram of the paths of load resistance in a structure is created.

Modern methods of analysis are based on increasingly precise computations, where efficiency is unnecessary because the computer is the primary calculating instrument. Because of the difficulty of computations in the late nineteenth century, methods from this period show an economy of calculation that could significantly benefit modern engineering. A few of these calculation methods are described in Chapter 5.

Sources of Information

The principal source of information for this book is the textbooks of the period. A very great number of very useful textbooks have been made available as free books on Google .com or HathiTrust.com. The most useful books have been the design manuals, such as Kidder's *Architects' and Builders' Pocket-Book*, or Trautwine's *The Civil Engineer's Pocket-Book*. Additionally, several catalogs have been consulted. Various other academic source materials are available and have been very useful to the development of this work. Engineering professors often mimeographed and bound their course notes, and some of these materials remain available in libraries throughout the country. Notable among these is George Fillmore Swain's notes, while the notes of Augustus Jay Du Bois and Charles Crandall also have been consulted. The records of the Berlin Iron Bridge Company, mostly available at the Huntington Library in San Marino, CA, also have been found to be very revealing of contemporary ideas of bridge and building design. Almost all of the published textbooks cited at the end of each chapter in this work are also available as free eBooks on Google .com or HathiTrust.com.

Many images have been obtained from the online material available in the Library of Congress's collection of Historic American Buildings Survey/Historic American Engineering Record (HABS/HAER) measured drawings and photographs. The catalog number is given in the caption for each of these images. The search box for this catalog can be found at http://www.loc.gov/pictures/collection/hh/.

Organization and Format of the Book

This book is divided into three major sections covering the three major types of design practiced in the nineteenth century: empirical, analytical, and graphical. Empirical rules for engineering fall generally into three classes. The first type of empirical rule is practice based, that is depending on precedent without further consideration. Contemporary examples of this type of rule include the application of span/depth rules. These methods particularly apply to the design of masonry arches, which is described in Chapter 2. A second class of empirical rule is a rational analysis that is abbreviated and used to develop rules to be applied to the design of specific structures. Examples of this practice are Hatfield's rules, described in Chapter 3. Chapter 4, describing the empirical design of metal structures, contains several results of column tests curve-fitted to the development of semiempirical formulas of the third class.

The following section of the book describes analytical procedures for design. Unlike the previous section, this section is divided by type of structure: the subject of Chapter 6 is the analysis of arches in masonry or iron and steel. Chapter 7 covers the analytical methods used for trusses in wood or iron, applied to building structures, highway bridges, and railroad bridges. The topic of Chapter 8 is analytical methods for the design of beams and girders, including continuous girders, whereas Chapter 10 describes the developed methods for the analysis of portal frames, which can be extended to more general frames.

Finally, the book describes the highly evolved methods of graphic analysis used during this time period. Chapter 11 is an introduction to graphical analysis to give the reader the opportunity to study the terms used and the general methods used in graphical analysis. The analysis method can be applied to arches, beams, and frames, and includes refined developments in geometry. Chapter 12 covers the graphical analysis of trusses, Chapter 13 is about the graphical analysis of arches, Chapter 14 concerns the graphical analysis of beams, and Chapter 15 describes the graphical analysis of portal frames and is comparable to the analytical methods presented in Chapter 10.

In the concluding Chapter 16, the influence of analysis and design methods on the design outcome is investigated. The remainder of the chapter consists of a case for the preservation of the methods of analysis of the late nineteenth century.

References Cited

American Association of State Highway and Transportation Officials (AASHTO). (2013). *AASHTO LRFD Bridge Design Specifications*, 6th Ed. Interim Revisions. AASHTO, Washington, DC.

American Institute of Steel Construction (AISC). (2011). *Steel construction manual*, 14th Ed. AISC, Chicago.

American Wood Council. (2006). ASD/LRFD, NDS, *National design specification for wood construction: With commentary and supplement*. American Forest and Paper Association, Washington, DC.

Waddell, J. A. L. (1894). *The designing of ordinary iron highway bridges,* 5th Ed. John Wiley and Sons, New York.

Part I

Empirical Methods

1

Empirical Structural Design

Throughout the past two millennia, two distinct currents of thought have guided the practitioners of the building arts: the empirical tradition and the scientific tradition. The empirical tradition in building is the application of skill and experience to the solution of problems: a carpenter who uses experience to size and erect floor framing is practicing empirical design. The scientific tradition, conversely, is the application of the results of speculative and experimental science to the practice of building. The application of Euler's formula $P_{cr} = \pi^2 EI/l^2$ to the design of columns is a common example of scientifically based design. Building practice has relied much more on the empirical approach throughout most of the nineteenth century and earlier. Even today, while purporting to design according to scientific principles, the engineering profession relies on the empirical approach to a greater degree than generally supposed.

Merriam-Webster's online dictionary defines the word *empirical* as "relying on experience or observation alone often without due regard for system and theory." On this basis, a significant body of engineering works in the United States is clearly designed empirically. Equally clear is that a large component of empirical design pervades the thinking of the contemporary engineering profession. The readiest example of empirical design is the choice of #4 @ 12 for a reinforcing bar size and spacing—a decision that is made without calculations for minor components of a concrete structure. In this instance, experience, not reason, dictates the configuration of the reinforcement.

Scientifically based design stands in opposition to the empirical tradition. Design decisions are based on analyzing the structure and understanding its response, using the laws of mechanics, strengthening by experiments, and proportioning members based on their expected response. Examples of this might be the analysis of a wood beam according to the Bernoulli-Euler theory of bending and the sizing of the beam based on limiting the maximum calculated bending stress. For reinforced concrete, scientific design would require determining the ultimate bending moment of a reinforced concrete slab (based on factored loading), the sizing of the reinforcement on the basis of its yield strength, and the moment arm of the reinforcement about the center of the compressive stress block on the compression face of the slab.

In the realm of built works, such as buildings and bridges, the application of empirical design must be considered alongside the application of scientific design. To begin with construction in ancient Rome, as seen through the Roman architect and writer M. Vitruvius Pollio, there is already a balanced application of scientific principles and the experience of builders to the design of public buildings, such as temples and basilicas (Granger 1931).

Vitruvius's principal mode of structural design is the development of rules for the proportioning of the elements of a building. His prescriptions for architecture involve applying rules of proportioning, or "symmetry" in the Vitruvian sense, to the production of works of architecture. Thus, columns are designed to shaft height/diameter ratios between 7 and 9½, depending on the order of the column (Figure 1-1). Although these ratios are partly intended to be pleasing to the eye, they are also an expression of a structural necessity. Where Vitruvius relates the intercolumniation to the height/diameter ratio, within the prescriptions of the Ionic order, he is suggesting the thickening of columns for larger spacing, both for the visual requirements of the building front and for the structural requirements of columns that will be called on to carry a greater load as well. More explicit is the suggestion that stone architraves (with a depth of 1/2 a column diameter) often break where an intercolumniation of 3 diameters is used, whereas for temples with an intercolumniation greater than 3 diameters, the architrave must be constructed of wood. For civil architecture, Vitruvius's prescriptions are similar. For the walls of occupied basements, the spacing of buttresses is to be the height of the wall, the buttresses are to taper from the projection at the bottom of the wall equal to the height of the wall to a projection equal to the thickness of the wall at the top, and the width of the buttresses is to be equal to the thickness of the wall (Granger 1931).

However, Vitruvius also participates in an important scientific tradition, inheriting some ideas from the Greek natural philosopher, Aristotle, and the adherents of his school. Aristotle taught that matter is composed of four elements: earth, water, fire, and air. Earth and water are heavy, and fire and air are light. A stone composed primarily of earth will be heavy but will have little resistance to moisture ingress. Fire, being light, will weaken a stone. These ideas can be found applied to building materials in Vitruvius's *Book II* (Granger 1931), which describes the properties of materials. Although Vitruvius's science seems erroneous, it enabled choices that a modern engineer would also make, for example, to allow freshly quarried stones to sit for a year to exude the weakening effect of the moisture, or to avoid stones subjected to fire that have been weakened by the addition of the element fire and should not be reused. Aristotle's followers also produced a short book titled *Mechanical Problems* (Hett 1936) that influenced Vitruvius and many later architects. In this work, most mechanical actions are based on the circle and its derivatives, the balance, and the lever. A lever is said to work because the effort applied to the longer arm moves with a greater velocity than the weight at the end of the shorter arm. Many other objects,

Figure 1-1. Column shaft diameter to height ratios, according to Vitruvius.
Source: From Leveil (n.d.).

including the beam, are explained in terms of the action of the lever. In the case of a beam, an external lever tends to open the beam about a fulcrum in the compression face and an internal lever within the beam counteracts the effect of the external lever. Motion, in the Aristotelian sense, may be natural: downward for heavy objects; or such motion may be forced or constrained by other agencies so that the object moves in other directions. Ideas about motion are applied to building objects, such as arches, columns, buttresses, and beams.

The medieval architects are known to have worked on the basis of established precedents and according to geometrical ratios. The greatest of the Gothic cathedrals resulted from the investigation and use of schemes of proportioning and the cautious increase of the size of buildings proportioned according to these schemes. Shown in Figure 1-2 is the section of the nave of Amiens Cathedral and some of the geometric logic that ensured the stability of that structure. Much of the attention of modern interpreters of medieval architecture has been on the construction of geometric diagrams representing the proportioning schemes evident in a given building. However, a few texts make it possible to infer the medieval mindset concerning structural design. The thirteenth-century drawings of Villard

Figure 1-2. Equilateral triangles superimposed on section of Amiens Cathedral.
Source: Image of the cathedral from Durand (1901).

de Honnecourt show various geometrical constructions applied to the layout of building structures. Other documents refer to buildings laid out in harmonic ratios, according to the square and the circle. Elevations of buildings are completed according to the square, as at Cologne, or the triangle, as at Amiens. The design of piers is based on proportioning as most of the major nave piers in Gothic buildings have a height/width ratio between 7 and 9. Most of the remaining documents concerning the design of medieval buildings refer explicitly to geometric ratios. However, a set of documents relating to the construction of the Cathedral of Milan from 1399 to 1401 is particularly revealing. (For a review of these documents, see Ackerman 1949.) In this case, the native engineers had to defend their design for the cathedral against the criticism of a visiting French architect, Jean Mignot. Mignot declared from his arrival onward that the structure was threatening ruin and appealed to the Duke of Milan. The reports of the various experts and council members are almost the only medieval construction documents in which theory is discussed, rather than a particular project, and are thus among the most studied records from medieval architecture. Mignot's list of 54 faults or "doubts" is presented on January 11, 1400, along with the responses of the Milanese architects. In a further council meeting, on January 25, Mignot elaborates on his main objections: the four towers intended to sustain the *tiburio* (crossing tower) are not built with sufficient foundation or piers, and the buttresses around the chevet are inadequate.

In the earlier meeting, in defense of their chevet scheme, the Milanese architects make the statement that "pointed arches do not exert a thrust on the buttresses." Having had some time to think about this, Mignot counters two weeks later:

> And what is worse, it has been rebutted that the science of geometry does not have a place in these matters, because craft is one thing and theory is another. The said master Jean says that craft without theoretical knowledge is worthless, and that whether vaults are round or pointed, if they don't have good foundation, they are nothing, and nevertheless when they are pointed, they have the greatest thrust and weight. (di Milano 1877; translation by author)

The Milanese architects, however, did insert iron tie rods to resist the thrust of the arch that they may not have understood well. The insertion of the iron tie rods can be interpreted as a gesture of empirical design, made by the Milanese engineers: with what little theoretical knowledge they may have displayed in the construction of the arches and vaults of the cathedral, they were aware of the horizontal thrust that they exerted on their supports, or the power of the iron tie rods to resist the horizontal thrust of the arches.

Fillippo Brunelleschi, celebrated as one of the first Renaissance architects, is renowned for his courage in vaulting the 72-braccia (144 ft, approximately) octagon prepared for the crossing of the Cathedral of Florence, the measures taken to relieve or redirect the forces from the weight of the dome, and especially for the machines that he invented for the construction of this dome. There is really very little science in Brunelleschi's activities; although he was partially educated, he was a particularly skilled mechanic and inventor (King 2000), and the greater part of his design for the dome at Florence must be characterized as empirical.

Conversely, Leon Battista Alberti made great contributions as an architectural theorist, as exemplified by his *Ten Books on Architecture* (Bonelli and Portoghesi 1966). In speaking of machines and structures, he sounds a much more practical note. *Ten Books* contains detailed descriptions of the functioning of machines, principally cranes. Alberti's discussions are consistent with the descriptions of such machines in Vitruvius. His view of weights is Aristotelian: "Loads are heavy by nature and obstinately search for the lowest point, and with all their power do not allow themselves to be raised" (*Book VI*, Chapter 6, Bonelli and Portoghesi, 1966, p. 477). By the art or ingenuity of men, according to Alberti, weights can be moved in different directions than their nature dictates.

Empirical design also was practiced widely in the nineteenth century. Although by this time scientific theories certainly had come to be applied regularly to the design of buildings, empirical rules and practical knowledge were a necessary adjunct to such design. Textbooks from the time contain a significant proportion of practical instruction, and course programs in the universities where civil engineering was taught also contain a large share of practical instruction. For instance, Baker (1907), in his *Treatise on Masonry Construction*, alternates practical and theoretical considerations. As an example of his practical mindset, in refuting Rankine's insistence on the middle third rule, Baker states, "A reasonable theory of the arch will not make a structure appear instable which shows every evidence of stability" (p. 451).

The conflict between differing theories about where the thrust line may lie in an arch, described in greater detail in Chapter 6, is a particularly good example of the conflict between theoretical ideas about arch behavior and empirical understanding of arch stability. The design of bridges in the nineteenth century was similarly composed of equal parts empirical knowledge and rational design. Although extremely sophisticated methods were applied to the design of masonry arches, such as the application of Méry's method to the design of the Union Arch, described in detail in Chapter 13, the determination of the configuration of these structures continued to be based on conventional ratios.

Another empirical builder of note, Rafael Guastavino, and later his son, produced clay tiles that were laid up flat in Portland cement to form vaults and domes in various configurations. Both the elder and the younger Guastavinos proposed in treatises that the domes their company built did not exert thrust on their supports. They argued that, being cohesive in nature and monolithic in character, their domes had an inherent resistance to bending, unlike voussoir arches and domes; therefore, they did not generate horizontal thrust. Although Rafael Guastavino the elder did resort to structural engineering arguments in explaining the action of his constructions, he was almost entirely an empirical builder, deciding on the number of layers of tile required by his vaults based on size and other considerations. However, many of the major structures designed and built by the company, such as the massive dome over the crossing of the Cathedral of St. John the Divine in New York, were built with iron reinforcing. When Rafael Guastavino the younger patented his system of construction, he showed probable locations for metal reinforcement (see Figure 1-3). The denial that their domes exerted thrust, coupled with their insertion of iron to resist the thrust, resembles the approach of the engineers at Milan, previously discussed. Any reader interested in a thorough study of the structural engineering achievements of Rafael Guastavino is directed to John Allen Ochsendorf (2010).

Wood structures have lent themselves to empirical design from the beginning of construction through the present. In the nineteenth century, many of these rules were codified and applied almost universally to design. Length-to-thickness ratios between 10:1 and 20:1 are usually applied to wood columns, whereas joists can reach span/depth ratios of 20:1.

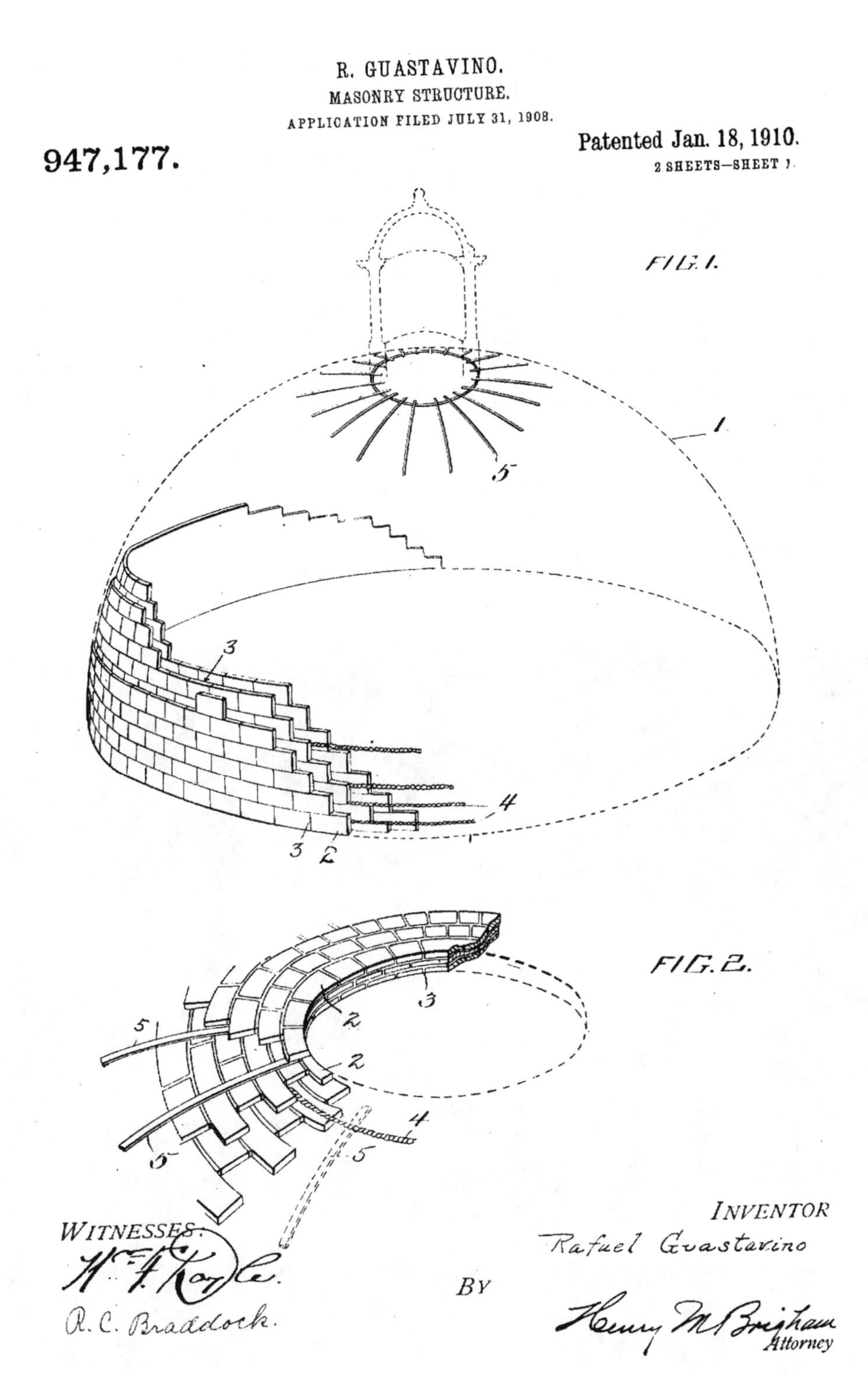

Figure 1-3. Guastavino's dome or vault.
Source: U.S. Patent Office, Patent No. 947,177, 1910.

Builders and architects adopted a multiplicity of basic configurations for iron bridges; many of them have commercial significance due to patents obtained on the design of the bridges. The two fundamental trends are effectively the arch and the truss. The arch, of course, follows the application of the stone arch and reproduces this form in cast iron. The oldest cast-iron bridge in the United States, James Finley's bridge at Brownsville, PA (Figure 1-4), is modeled on a stone arch. Other early examples of iron arch bridges include the patented design for iron arches of Thomas Moseley (Figure 1-5). The truss form was relatively slow to develop into the familiar assembly of smaller pieces into a single load-bearing structure. Early trusses were more experiments in bracing a longer top and bottom chord. Bow's methods, which evolved into methods for the analysis of trusses, were originally intended as analysis methods for braced beams. The web members were thought of as bracing for the remainder of the structure, either an arch, as in Moseley's designs, or a beam, as in a queen-post or a Howe truss.

Many of the bridge forms used by engineers of this time led to statically indeterminate structures, especially those types of bridges that had multiple web systems, for instance, the double Pratt truss (Whipple truss), the double Warren truss, and the multiple Warren truss, as shown in the examples of the Hayden Bridge, OR (Figure 1-6), the Sugar Creek Bridge

Figure 1-4. Dunlap's Creek Bridge (1839), Brownsville, PA.
Source: Photograph by the author.

Figure 1-5. Arched wrought iron bridge by Thomas Moseley. Upper Pacific Mills Bridge moved to Merrimack College, North Andover, MA (HAER MASS,5-,LAWR,6-).
Source: Photograph by Martin Stupich.

near Troy, PA (Figure 1-7), and the Slate Run Bridge in Slate Run, PA (Figure 1-8), respectively. The statically indeterminate aspect of these bridges was managed through an empirical procedure of dividing the bridge into multiple systems and analyzing each of the systems separately.

The knowledge developed by bridge engineers and incorporated into their designs and textbooks went well beyond the application of methods for stress analysis in the chords of the trusses. It included careful adaptation of details to various conditions and a willingness to allow statical indeterminacy in the design of bridges by permitting approximate analysis. The adoption of conventional bridge forms and the application of these forms to nearly all bridges of spans of 200 ft or less amount to a form of empirical design. In the end, such bridges were eventually built almost exclusively as through Pratt trusses.

Column design in this period is based on a semiempirical understanding of column behavior, bolstered by a few widely publicized experiments. It was certainly understood that buckling reduced the apparent strength of longer columns, but consistent means for measuring the strength of a column were not used. For steel columns, various competing formulas for the reduction of the strength of a column were applied with different factors, based on

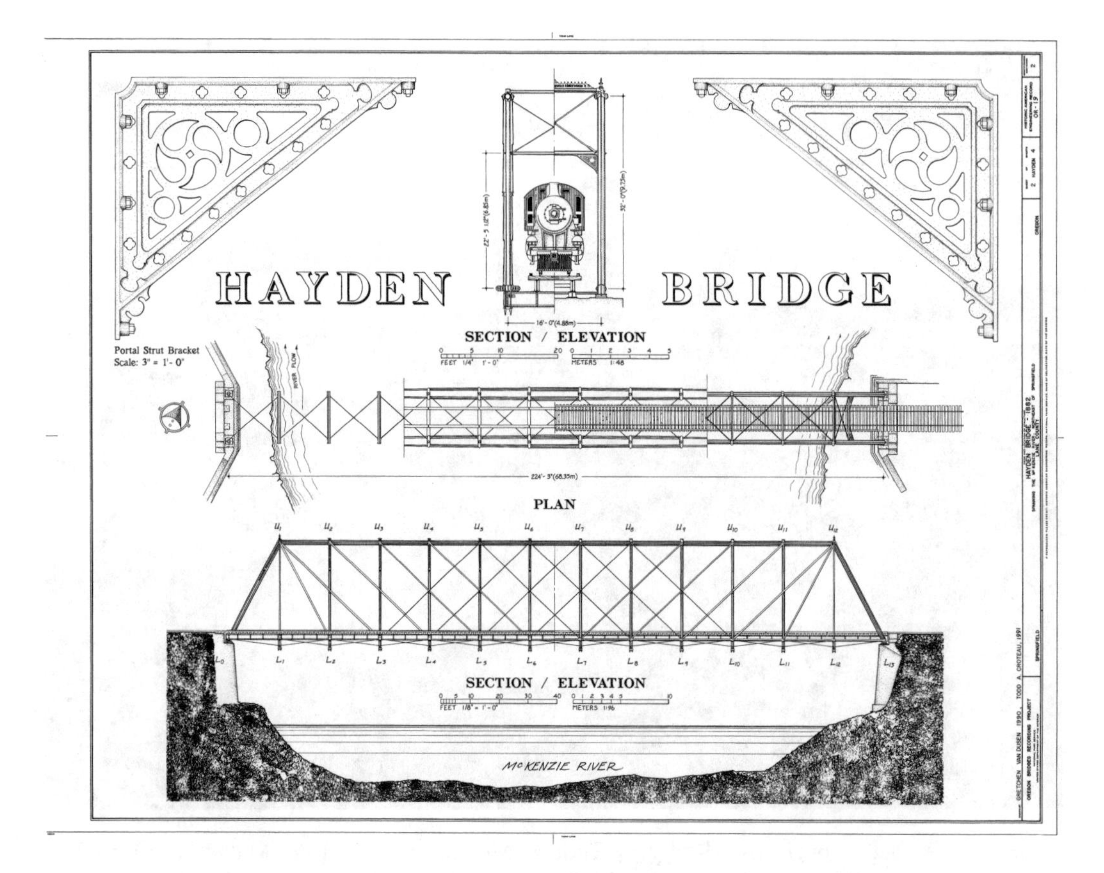

Figure 1-6. Hayden Bridge spanning McKenzie River at Southern Pacific Railroad (moved from Springfield, Lake County, OR), Springfield, Lane County, OR. Double-intersection Pratt truss (HAER ORE, 20-SPRIF, 2-).
Source: Historic American Engineering Record.

the nature of the column cross section. For timber columns, strictly empirical formulas for the reduction of strength were applied.

In the final analysis, structural design of bridges and buildings in the nineteenth century contains significant elements of both empirical design and rational/scientific design. This is more the nature of structural design than a temporary condition, in which the empirical elements will be overcome by the rationality of the scientific method. The actions of structures are too complex to examine in detail scientifically; the understanding of the producers and the users of the structures depends so much on the visual associations produced by a conventional structure that it is necessary to add a significant part of collective building experience into the design of the simplest and the most complex of structures.

At present, empirical design is widely used by engineers in several basic forms. The first form is the use of empirical rules, such as span/depth ratios for the proportioning of structural elements. Although this practice usually is implicit in a designer's selection of the preliminary size of an element for design, at least the building code for reinforced design makes the use of span/depth ratios explicit. Design professionals in structural engineering

Figure 1-7. Double Warren truss bridge (1907), Bronson Road over Sugar Creek, near Troy, PA.
Source: Photograph by the author.

also use invariant bay sizes for certain building uses and for certain structural materials. The design of bridges is still ruled by empirically based ratios: slab depth/span ratios, although not codified, are firmly established in the minds of bridge designers. Distribution factors, however, are embodied in the design code; these factors are critical in determining the live load distributed to each girder. The determination of these factors was, until recently, strictly empirical, being based on arbitrary ratios of the girder spacing to a constant (called S-over factors). In recent bridge design codes, the distribution of loads to girders is based on an empirical multivariate formula to determine the number of wheel loads to be distributed to each girder (AASHTO 2012).

A more widespread application of empirical design is the insistence of structural engineers in contemporary practice on using methods for design that are better justified by precedent than by any form of rational analysis. An example of this type of design is a one-way slab. In the usual design procedure, the slab is assumed to be simply supported and uniformly loaded for the calculation of bending moments; furthermore, a unit strip is assumed to behave the same as the slab, and the slab is assumed to be reinforced at mid-depth. None of

Figure 1-8. Upper Bridge at Slate Run, spanning Pine Creek at State Route 414, Slate Run, Lycoming County, PA. Quintuple Warren truss bridge (HAER PA,41-SLARU.V,1–3).
Source: Photograph by Joseph Eliot.

these assumptions are justifiable on the basis of carefully observed slab behavior. Surely the slab has multiple spans, and surely the reinforcement will sink to the bottom throughout most of the slab. The use of certain minimum values not necessarily embraced by the building codes, in a variety of situations, also qualifies as empirical design. Examples are sidewalk control joints every 5 ft, a minimum slab on grade thickness of 4 in., and a great variety of other procedures that are indispensable for effective practice of engineering.

Structural engineers, especially following the building codes that have appeared over the past century, often find themselves applying formulas for which it is impossible to see the rational basis. This is a form of empirical design in which the analysis that precedes the design has become so complicated or cumbersome that the design is ultimately based on ignorance of the principles used in the analysis.

In spite of their seeming irrationality, the methods practiced by all the designers described in this chapter, empirical or scientific, have a convincing justification: they work. James Ackerman's words about the serious errors in the fourteenth-century design of the Cathedral of Milan are appropriate here:

> Time and again northern masters expose the inadequacy of the entire structural system, attribute to it faults of the greatest magnitude, and leave, convinced that the work is destined to ruin. The Milanese plod stubbornly along ... determined to accept no foreign solutions to the major problems in construction. ... Only one argument, and an incontrovertible one, speaks in favor of the Milanese: the Cathedral was built entirely according to their designs and it stands (1949, p. 104).

References Cited

Ackerman, J. (1949). "'Ars sine scientia nihil est.' Gothic architecture at the cathedral of Milan." *Art Bulletin*, 31(2), 84–111 (in Italian).

American Association of State Highway and Transportation Officials (AASHTO). (2012). *AASHTO LRFD design specifications*. AASHTO, Washington, DC.

Annali della fabbrica del duomo di Milano. (1877). G. Brigola, Milan.

Baker, I. O. (1907). *A treatise on masonry construction*, 9th Ed. John Wiley and Sons, New York.

Bonelli, R., and Portoghesi, P., eds. (1966). *Trattati di architettura*. Edizioni il Polifilo, Milan (in Italian).

Durand, G. (1901–1903). *Monographie de l'église Notre-Dame, cathédrale d'Amiens*. Yvert et Tellier, Amiens, France (in French).

Granger, F. (1931). *Vitruvius on architecture*. Loeb Classical Library, Cambridge, MA.

Hett, W. S. (1936). *Mechanical problems*. In *Aristotle: Minor works*. Loeb Classical Library, Cambridge, MA.

King, R. (2000). *Brunelleschi's dome*. Walker, New York.

Leveil, J. A. (n.d.). *Vignole, traité elémentaire pratique d'architecture*, Nouvelle édition. Garnier Frères, Paris (in French).

Ochsendorf, J. A. (2010). *Guastavino vaulting: The art of structural tile*. Princeton Architectural Press, New York.

2

Empirical Design of Masonry Structures: Brick, Stone, and Concrete

This chapter is concerned with the application of empirical rules to the design of masonry structures. Such rules relate the depth or thickness of the masonry in an element, such as an arch, vault, or wall, to the conditions of span, radius, geometry, or material that dictate the thickness requirement. In the empirical design of masonry, the determination of the size of the wall, arch, vault, or buttress depends only on the application of geometrical ratios and has little regard for loading, stress, forces, or other conditions. Empirical techniques were widely applied to bridge structures in the nineteenth century, according to rules promulgated by various authors, such as William John Macquorn Rankine and John Trautwine, among others. Similar rules were applied to the design of arches in buildings. The thickness of walls in buildings generally was determined in relation to the overall height of the wall supported.

A typical masonry arch bridge, as illustrated in Figure 2-1, has as its principal load-carrying component a barrel vault or "arch barrel." The arch barrel consists of an arch face of carefully constructed masonry on the visible sides of the bridge, often with rougher sheeting between the two faces. The strength of a masonry arch bridge is well established. As long as the supports of the arch remain fixed in the horizontal direction, the arch resists vertical loads by the development of horizontal and vertical internal forces that tend to follow the shape of the arch barrel. The self-weight of the arch and of the fill above the arch tends to introduce greater axial compression into the arch barrel that prestresses the arch and increases its resistance to moments induced

Figure 2-1. Pithole stone arch bridge, spanning Pithole Creek at Eagle Rock Road (State Route 1004), Pithole City, Venango County, PA (HAER PA,61-PIT,1-).
Source: Photograph by Joseph Eliot.

by axle loads or other concentrated loads. At the supports, the arch has both horizontal and vertical reactions, which must be resisted by the foundation system.

A level road surface is provided by the construction of vertical spandrel walls above the arch barrel. Fill is placed between the spandrel walls so the spandrel walls are primarily earth-retaining structures. As such, the spandrel walls are subjected to transverse soil pressures, often including the effect of surcharge due to loading on the roadway surface. Wing walls, either in plane with the spandrels or angled (as shown in Figure 2-2), are provided at the ends of the structure, as the lower grade is brought up to the level of the roadway. A parapet usually is provided above the level of the road surface on road bridges.

Rankine's and Trautwine's Formulas

The foundation of an arch bridge consists of abutments at the ends of the bridge and of internal piers between spans. The abutment must resist the horizontal and vertical reactions

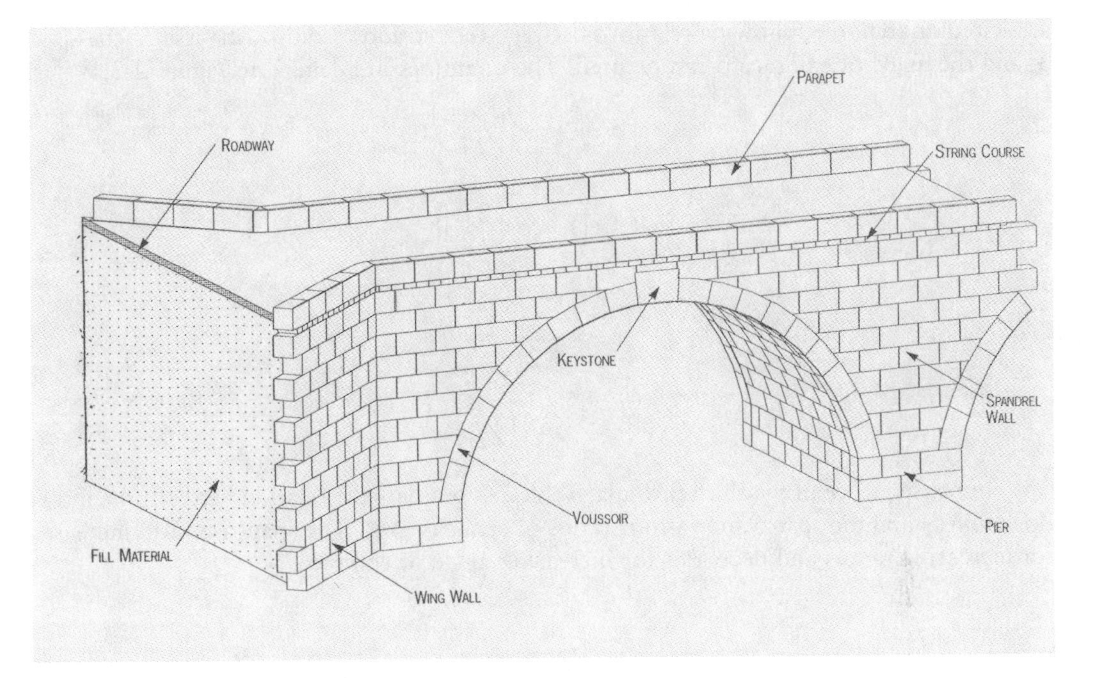

Figure 2-2. Masonry arch bridge nomenclature.
Source: Drawing by Carmen Gerdes.

due to the weight of the arch and to superimposed loads. At an internal pier, however, the resultant horizontal reaction from two adjacent arches of similar spans tends to be in equilibrium and to impose a relatively small horizontal force component on the pier. Although graphical methods for analysis of arches are widespread by the late nineteenth century (graphical methods of arch analysis will be reviewed in Chapter 13), some authors are more interested in the development of analytical procedures for the determination of the forces in an arch. Rankine (1865) appears to be foremost among those authors favoring an analytical approach, although many of the other English and some of the French authorities advocate similarly. Conversely, the application of analytical formulas is not indicated for most projects, even according to Rankine. Instead, it is customary to announce highly simplified rules for the calculation of the thickness of the arch at the crown and at the abutments and to work with simple rules for tapering the arch from the joint of rupture near the abutment to the crown. Rankine himself, having devoted dozens of pages to the development of analytical formulas for the arch (discussed in Chapter 6), gives one or two rules for the thickness of the arch, specifically intended to be employed in practice. A similar intent is evident in E. Sherman Gould's paper, "Proportions of Arches from French Practice" (1883). Based on simple observation of the size of arches, he deduces empirical rules for the proportioning of arches.

The comparison of these formulas and the interpretation of design methods for arch bridges in general require conversion equations between the important geometric quantities for an arch bridge. To determine the angle of embrace or intrados radius for a segmental or semicircular arch, the following relations between the intrados radius r, the rise R, the span S, and the angle of embrace β can be used. The quantities are defined in Figure 2-3.

$$S = 2r\sin\left(\beta/2\right)$$

$$R = r\left(1 - \cos\left[\beta/2\right]\right)$$

$$r = \frac{R^2 + \left(S/2\right)^2}{2R}$$

$$\beta = 2\sin^{-1}\left(S/2r\right)$$

In consequence of the third formula, Table 2-1 presents the relation between the intrados radius r and the span S for various ratios of span/rise S/R. The span/rise ratio increases for increasing radius and decreases for increasing angle of embrace.

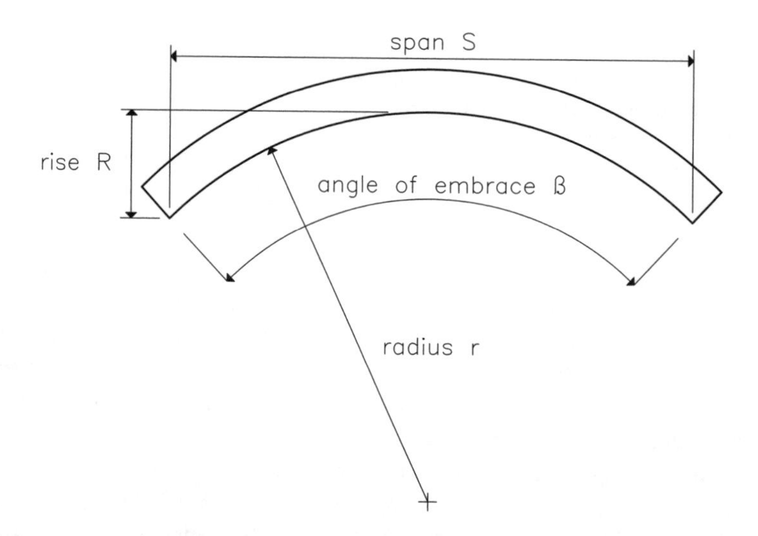

Figure 2-3. Geometry of a masonry arch.
Source: Author-provided figure.

TABLE 2-1. *Examples of Span, Rise, and Angle of Embrace Geometric Quantities*

S/R	*r/S*	β
2	0.50	180°
4	0.56	63°
6	0.83	37°
8	1.06	28°
10	1.30	23°

Rankine's (1865) rules were developed based on the variation in arch pressures for arches of varying span and rise and the finding that these pressures vary approximately according to the radius of the arch; hence, the depth must vary according to the square root of the radius. The coefficients are empirically based, although Rankine provides a theoretical justification for the form of the coefficients based on previously established calculations of the force in an arch. In his own words,

> To determine with precision the depth required for the keystone of an arch by direct deduction from the principles of stability and strength would be an almost impracticable problem from its complexity. That depth is always many times greater than the depth necessary to resist the direct crushing action of the thrust. The proportion in which it is so in some of the best existing examples has been calculated, and found to range from 3 to 70. The smaller of these factors may be held to err on the side of boldness, and the latter on the side of caution; good medium values are those ranging from 20 to 40. The best course in practice is to assume a depth for the keystone according to an empirical rule founded on dimensions of good existing examples of bridges (p. 290).

Rankine's rule for the thickness of the arch at the crown is as follows:

> For the depth of the keystone, take a mean proportional between the radius of curvature of the intrados at the crown, and a constant, whose values are,
>
> For a single arch.. 0.12 ft
> For an arch forming one of a series................................. 0.17 ft. (p. 425)

Whereas other authors assert a minimum value and a linear relationship between span or radius and ring thickness, Rankine considers a strict relationship of the depth of the keystone to the square root of the radius, expressed in more modern terms as

$$t_{\text{keystone}} = 0.35\, r^{1/2}, \text{ for a single arch and}$$

$$t_{\text{keystone}} = 0.41\, r^{1/2}, \text{ for a multiple arch.}$$

The input and output units in these formulas are feet.

Rankine justifies the square root relationship between arch depth and radius by several simplifications to a general formula he has developed for the tension in an arch loaded critically by a rolling load. Based on assumptions about the ratio between dead and live load and the span of the arch between joints of rupture, he arrives at a law similar to the one as shown, which varies depending on dead load/live load ratio and rise/radius ratio. Numerically, for conventional cases, the values presented as shown appear to be average results of this formula.

Trautwine's (1874) rule is

$$t_{\text{keystone}} = 0.25(r + s/2)^{1/2} + 0.2$$

where r is the intrados radius, s is the span, and all input and output units are feet. The depth is to be increased by one-eighth for second-class work and by one-third for rubble or brickwork arches. Trautwine was a proponent of the practical, as opposed to the mathematical. In his opening comments, Trautwine claims that the writings of Rankine, Henry Moseley, and Julius Weisbach are beyond the reach of ordinary engineers, and he further asserts that simple facts can be "buried out of sight under heaps of mathematical rubbish" (p. viii). Trautwine's derivation of empirical rules, which results from completing a series of designs based on a parametric study of stone bridges, is representative of this viewpoint.

Rules Derived from French Practice

In keeping with Trautwine's outlook, Gould (1883) developed a series of rules by a combination of observation of standing bridges and review of the writings of eminent French engineers. Jean-Rodolphe Perronet's rule, endorsed by P. Léveillé, is to make the thickness of the arch at the crown e, in feet (Perronet would have originally stated his rule in feet), such that

$$e = 1 + 0.035\ S$$

where S is the span in feet. Although Perronnet's rules, like Rankine's and Trautwines's, are meant to cover all arch types, Julien Dejardin advances different rules for different arch forms, so

for rise/span = 1/2 (semicircular)	$e = 1 + 0.10R$ (effectively 0.05 S)
for rise/span = 1/6	$e = 1 + 0.05\ R$
for rise/span = 1/8	$e = 1 + 0.035\ R$
for rise/span = 1/10	$e = 1 + 0.020\ R$

where R is the radius of the intrados. For elliptical, or false-elliptical (three-centered) arches,

for rise/span = 1/3	$e = 1 + 0.07\ R$

Gould adopts the simplest and most all-embracing formula, Perronet's, "because the results it gives tally so well with many existing structures" (p. 451). He adds a modification of 2% of the fill height, where the fill height at the crown exceeds 2 ft. He presents several formulas for the increase in the ring thickness from the crown to the joint of rupture, including the commonly used projected vertical area formula, described following, and sets the joint of rupture arbitrarily at 30 deg from the horizontal. Thus, as shown in Figure 2-4, all that is required for the layout of an arch are Perronet's rule or some other rule for determining the depth (often called thickness) of the keystone at the crown; a rule for locating the joint of rupture, which is effectively the beginning of the abutment; and, finally, a rule for increasing the depth of the arch ring from the crown to the abutment.

Figure 2-4, from Gould's article, describes the empirical layout of a circular arch. The joint of rupture is located at 30 deg from horizontal. At this angle, the length of the joint is twice its vertically projected area, so twice the thickness at the crown, deduced from Perronet's formula. The remainder of the extrados is plotted according to Gould's interpretation of J. Dubosque: finding the bisector of the line joining the two known points of the extrados and plotting the circular arc that joins these two points. The bridge in the Allegheny

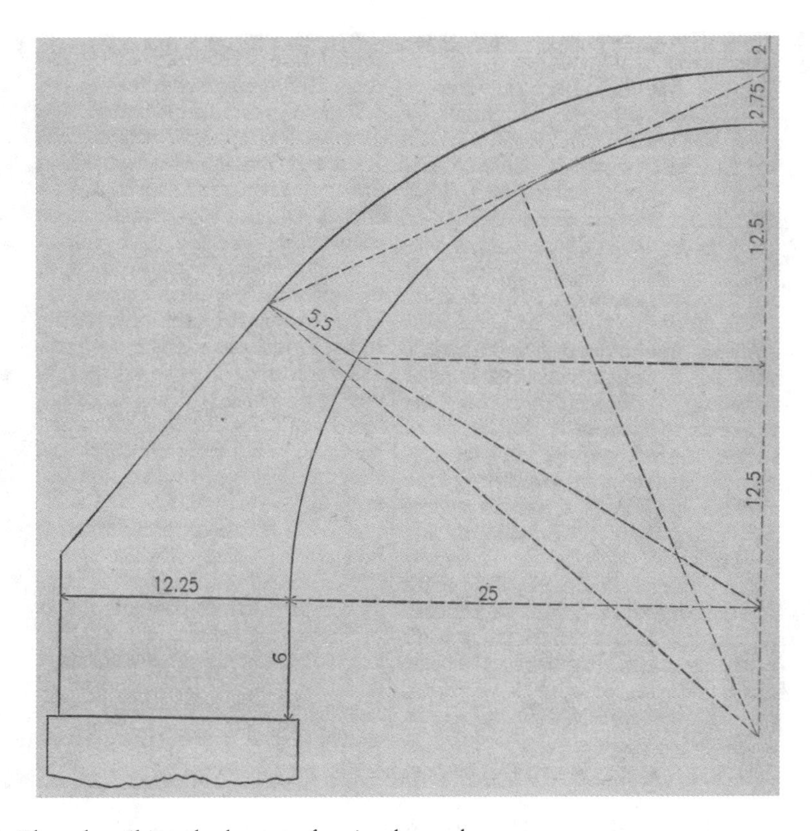

Figure 2-4. Plate describing the layout of a circular arch.
Source: Gould (1883).

Courthouse and Jail was designed by architect Henry Hobson Richardson and modeled after the Bridge of Sighs in Venice. Unlike its Venetian model, the Pittsburgh bridge has an exposed arch ring that follows rules similar to Rankine's and Trautwine's rules (see Figure 2-5). The depth of the voussoirs varies, the depth at the abutment having a vertical projection approximately equal to the depth at the crown, similar to many of the aforementioned rules. The depth at the crown is somewhat larger than prescribed by either Rankine's or Trautwine's rules.

Although he presents all these formulas, Ira Osborn Baker is dubious concerning their merit (1907, p. 495). These formulas are derived by the investigation of actual structures, but Baker comments that the structures may have been chosen because they agreed with the formula, rendering this comparison meaningless, and that while the safety of the structures may be inferred, nothing is known about the degree of safety. He also expresses concern about the lack of consideration for differences in materials of construction, differences in loading, differences in fill height, and other factors. "Many masonry arches are excessively strong; and hence there are empirical formulas which agree with existing structures, but which differ from each other 300 or 400 per cent" (p. 495). The table of comparison Baker constructed for a few normative cases of bridge construction is reproduced in Table 2-2.

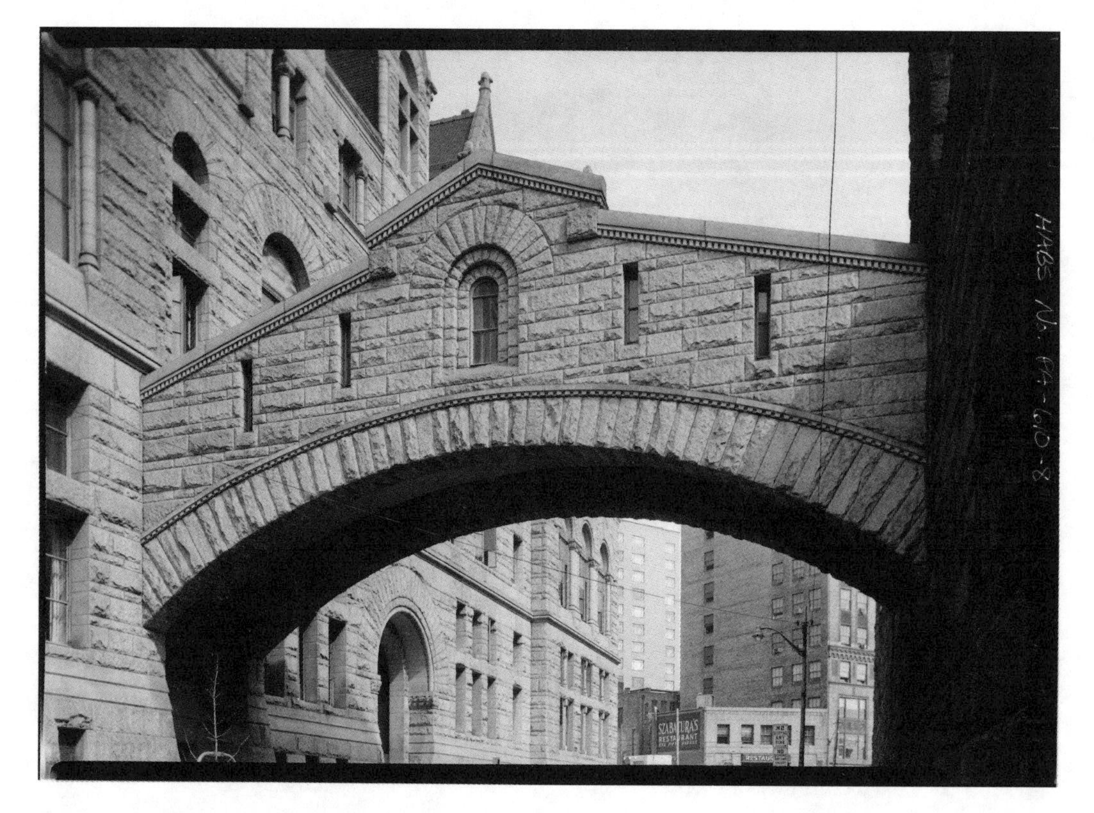

Figure 2-5. The Bridge of Sighs from H.H. Richardson's Allegheny Courthouse and Jail, Pittsburgh, PA (1884-6) (HABS PA,2-PITBU,29–8).
Source: Photograph by Jack E. Boucher.

TABLE 2-2. *Comparisons of Calculations of Depth of Arch Ring at Crown for Various Bridge Geometries*

	Proportion of Rise to Span								
	Semicircle			*Rise/Span = 1/6*			*Rise/Span = 1/12*		
	Span			Span			Span		
Formula	*10*	*50*	*100*	*10*	*50*	*100*	*10*	*50*	*100*
Trautwine's	0.99	1.98	2.70	1.11	2.23	3.09	1.26	2.57	3.55
Rankine's	0.77	1.73	2.45	1.00	2.25	3.16	1.25	2.79	3.95
Perronnet's	1.51	3.26	5.43	1.51	3.26	5.43	1.51	3.26	5.43
Dejardin's	1.50	3.50	6.00	1.42	3.07	5.17	1.26	2.30	2.60
Croizette-Desnoyer's	1.38	2.48	3.30	1.56	2.86	3.85	1.62	3.01	4.05

In addition to the rules for the depth of the arch ring at the key, Rankine and other authors describe the tapering of the arch from the abutment to the crown. Because mechanics indicates a greater thrust at the abutment (horizontal and vertical force components) than at the crown (horizontal force component only), the arch barrel should become thicker as it approaches the abutment. Most authors state the simplest rule for this thickening, which is that the vertically projected area of each joint should be the same; that is, if φ is the angle of variation of the joints in the arch ring from the vertical and t is the thickness of the arch at the crown, the thickness at any other part of the arch should be $t/\cos\varphi$.

According to Rankine, the limits of the abutment, and of the haunching that is placed behind the arch barrel for stability, are the location of the *joint of rupture*. To an analyst, as will be seen in Chapter 5, the joint of rupture is the point of tangency of the thrust line with the intrados of the arch. However, in an empirically designed arch, the joint of rupture is the limit above which centering is required during the construction of the arch (Gould 1883, p. 453).

Abutment Rules

In addition to the aforementioned development of the thickness of the arch ring at the abutment, Rankine presents specific formulas for the size of abutments for arch bridges. Rankine's method is semiempirical; that is, calculating the center of pressure of the combined weight and lateral thrust on the abutment is required. The location of the center of pressure (on a horizontal plane at the base of the abutment) is expressed as a fraction q; for instance, for a rectangular pier q is limited to one-sixth of the thickness to ensure that the entire bearing is in contact. Rankine presents a target value of q based on observations on successfully built buttresses.

For interior piers, Rankine (1865, p. 428) proposes to determine the unbalanced lateral load by an empirical formula: multiply traveling (i.e., live) load per lineal foot × radius on intrados at the crown. Thus, the pier on a semicircular arch of span 50 ft, subjected to a load of 600 lb/ft (an estimate of a train load), would have to support an unbalanced lateral load of 15,000 lbs. By statics, the horizontal thrust due to live load could be no more than one-half of this quantity, in this case. For flatter arches, this proportion is greater. In practice, Rankine states that piers are constructed at least one-tenth of the span in width and more commonly one-sixth or one-seventh of the span.

Retaining Wall, Buttress, and Building Arch Rules

Trautwine (1874, p. 331) recommends height/base ratios ranging from 2 to 4 for rectangular retaining walls. He takes pains to point out that a retaining wall can be safely battered without diminishing its resistance. Similar to these considerations, Rankine states that according to the practice of British engineers, a value of q, the center of pressure through the base of the wall of up to 0.375, is tolerated, whereas the French engineers limit this value to 0.30–0.25.

Kidder's *Architects' and Builders' Pocket-Book* (1886, p. 187) describes empirical design procedures for buttresses, arches, and foundations for buildings. For the radius of arches over doors, windows, and other small openings in buildings, Kidder proposes a segmental arch with radius equal to width of opening. Kidder quotes Rankine's and Trautwine's rules and presents a table based on Trautwine's rule. Rankine's rule, having no minimum thickness, is virtually unworkable for short-span arches (3-ft arch span yields keystone depth of 2.4 in.); it is really meant to be applied to bridges. Kidder notes that both formulas produce unstable arches for a proposed 20-ft-span arch and follows with a discussion of graphic analysis of arches. He quotes a rule from the New York City building code, where a thickness of approximately 4 in./4 ft of span is required. His procedure for the design of buttresses is strictly graphic and will be described in Chapter 13. For foundation work, the thickness of foundation walls is guided by experience and generally embodied into building codes, such as Kidder's description of the building laws of New York and Boston. Footings are made slightly broader than the wall or pier that they support but are limited by the rule of having no back joints beyond the line of the wall or pier supported (Kidder, p. 143), unless the walls are in double courses (some of his diagrams violate this rule). The walls, at least according to the Boston building code of the time, must project a minimum of 12 in. beyond the supported element. A thorough reading of the New York, Boston, or Chicago building codes would enable the design of the entire foundation system for a conventional building.

Similarly, the Chicago building code (Chicago City Council 1905) gives explicit guidance on the required thickness for the exterior and interior walls of various classes of buildings: office, commercial, hotel, theater, and so on. An example of a four-story mercantile building is shown in Figure 2-6. Floor-to-floor heights of 10 to 12 ft are assumed in most cases. The exception is the requirements for theaters, where the thickness of the wall is dependent on the unsupported height. For walls in commercial buildings up to 25 ft high, a thickness of 20 in. is required, resulting in a maximum height/thickness ratio of 15, whereas a wall up to 90 ft high is required be 30 in. thick, for a height/thickness ratio of 36. The Monadnock Block in Chicago, shown in Figure 2-7, has walls said to be 6 ft thick at the base. The wall thickness is reduced every third floor. Although the walls are somewhat thicker than prescribed by the Chicago building code, the walls are perforated at the base, which may have induced the architects to use thicker walls.

The rules in the Cleveland building code (City of Cleveland, Ohio 1904) are presented in the graphic form shown in Figure 2-8. The result of these rules is load-bearing walls of similar thickness to those prescribed by the Chicago building code.

Although the empirical methods presented here are suitable for most types of bridge and building arches, and were probably used for the overwhelming majority of such arches, it was occasionally felt that a more analytical procedure may yield a better result. In such cases, the designers could have recourse to analytical procedures, such as those described in Chapter 6, or to graphical procedures, such as presented in Chapter 13. In the design of buildings, however, it is unlikely that designers would have used any but empirical methods for the design of arches, and, similarly, load-bearing walls were almost exclusively designed by empirical methods.

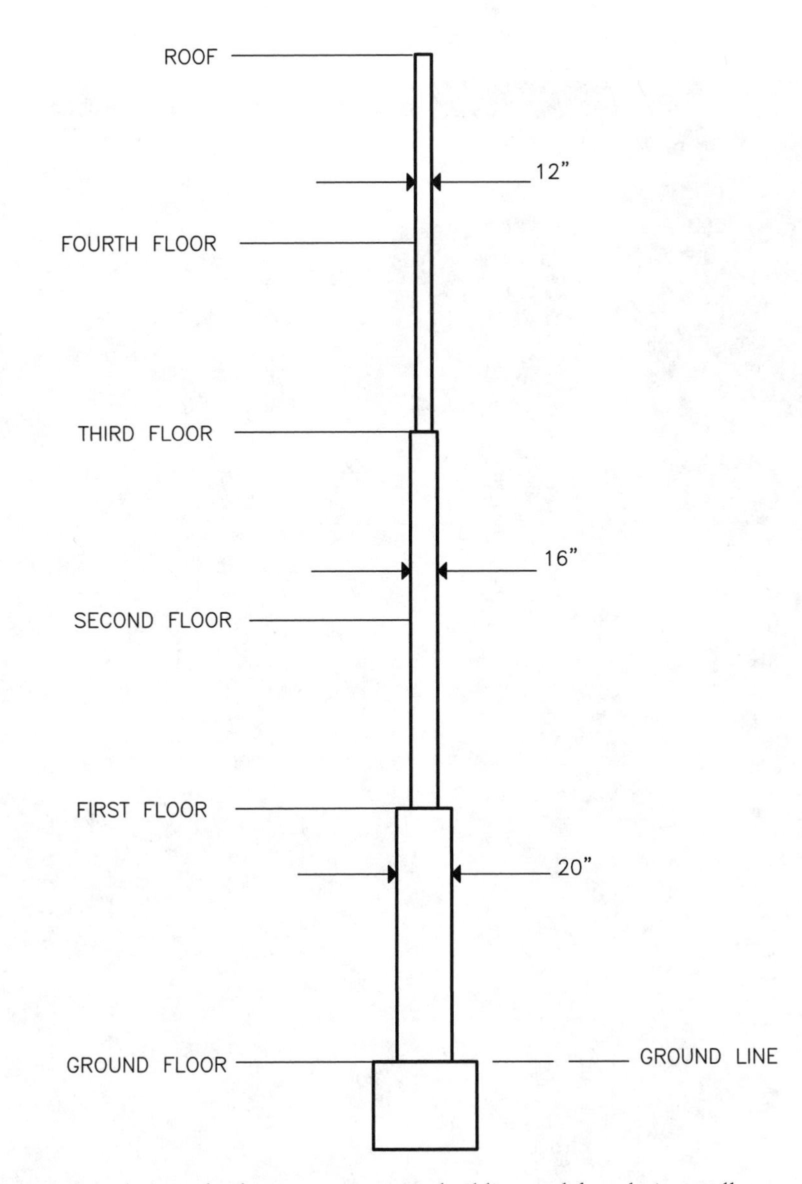

Figure 2-6. Explicit design of a four-story masonry building and foundation wall.
Source: Chicago City Council (1905, sect. 101).

Figure 2-7. Photograph of Monadnock block, Chicago (1891) (HABS ILL,16-CHIG,88-). The walls, almost 6 ft wide at the base, taper upward, changing every second level, in the manner of the diagram in Figure 2-6.

Source: Photograph by Cervin Robinson.

TABLE-B

I	II	III	IV	V	VI	I
Height of Building in Stories	Thickness of Wall in Brick		Height of Wall in Feet Measured Downward From ⊙ Top of Coping ⊙ To	Thickness of Wall in Brick		Height of Building in Stories
	Floor Load not over 60 Lbs. per Sq. Ft.			Floor Load 60 Lbs. and over per Sq. Ft.		
Parapet Wall	1½	1½	3 Roof Line 3	1½	1½	Parapet Wall
Attic or Air Space	1½	1½	Min. 6 to Max. 10	1½	1½	Attic or Air Space
One	1½	1½	18 " 24	1½	1½	One
Two	1½	1½	30 " 38	1½	2	Two
Three	1½	2	42 " 52	2	2	Three
Four	2	2	54 " 66	2	2	Four
Five	2	2	66 " 80	2	2½	Five
Six	2	2	78 " 94	2½	2½	Six
Seven	2	2½	90 " 108	2½	3	Seven
Eight	2½	2½	102 " 122	2½	3	Eight
Nine	2½	2½	114 " 136	3	3	Nine
Ten	2½	3	126 " 150	3	3½	Ten
Eleven	3	3	138 " 164	3½	3½	Eleven
Twelve	3	3½	150 " 178	3½	4	Twelve
Thirteen	3½	3½	162 " 192	4	4	Thirteen
Fourteen	3½	4	174 " 200	4	4½	Fourteen
Fifteen	4	4	186 " 200	4½	4½	Fifteen
Sixteen	4	4½	198 200	4½	5	Sixteen
Length of Walls	Up to 60 Ft.	Above 60 Ft. Unlim.	To Base of Wall Story per Story	Up to 60 Ft.	Above 60 Ft. Unlim.	Length of Walls
Basement Walls	Shall be one-half (½) brick thicker than wall next above					Basement Walls

Figure 2-8. Table B prescribing the thickness of brick masonry bearing walls for buildings of various heights.
Source: City of Cleveland, Ohio (1904).

References Cited

Baker, I. O. (1907). *A treatise on masonry construction*, 9th Ed. John Wiley and Sons, New York.

Chicago City Council. (1905). *An ordinance relating to the department of buildings and governing the erection of buildings, etc.* Moorman and Geller, Chicago.

City of Cleveland, Ohio. (1904). *The building code of the City of Cleveland, Ohio.*

Gould, E. S. (1883). "Proportions of arches from French practice." *Van Nostrand's engineering magazine*, 29, 449–469.

Kidder, F. (1886). *The architects' and builders' pocket-book*, 3rd Ed. John Wiley and Sons, New York.

Rankine, W. J. M. (1865). *A manual of civil engineering*, 4th Ed. C. Griffin, London.

Trautwine, J. C. (1874). *The civil engineer's pocket-book.* Claxton, Remsen, and Haffelfinger, Philadelphia.

3

Empirical Design of Wood Structures

In the latter part of the nineteenth century, wood was a commonly used structural material owing to its wide availability in all regions of the country. Significantly larger sizes of timber and significantly longer timbers were available compared with current-day use of the material, although limitations in transportation may have restricted the number of species available. The Chicago building code 1905 (Chicago City Council) listed three basic types of wood: spruce or white pine, presumably from Wisconsin or Minnesota; loblolly yellow pine, probably from southeastern U.S. forests; and white oak, from the eastern forests of Ohio or Pennsylvania. Wood construction generally was done by skilled carpenters, often with little engineering input, using either standard practice or rules of thumb. The framed roof of the 1889 Altoona, PA, Masonic Temple, from Historic American Buildings Survey (HABS) records (Figure 3-1), shows the use of trusses and joists. The trusses include a significant amount of joinery in the connections between wood elements, and the joists appear to be toenailed to the supporting trusses. The roof sheathing consists of boards. A large amount of joinery also is shown in Robert Griffith Hatfield's (1895) image of a framed opening for a chimney (Figure 3-2), where each element is mortised into the supporting element: header to trimmers, tail joists to header. This is typical of earlier carpentry applications and indicative of the skill of the carpenters and of the lack of standard fastenings. In applications ranging from residential to industrial, wood posts were widely used for the support of floors, whereas trusses were used to frame longer floor and roof spans. The design of all these features was often left to experience or to the application of derived rules, both of which methods are described in this chapter. Both these methods are instances of empirical design.

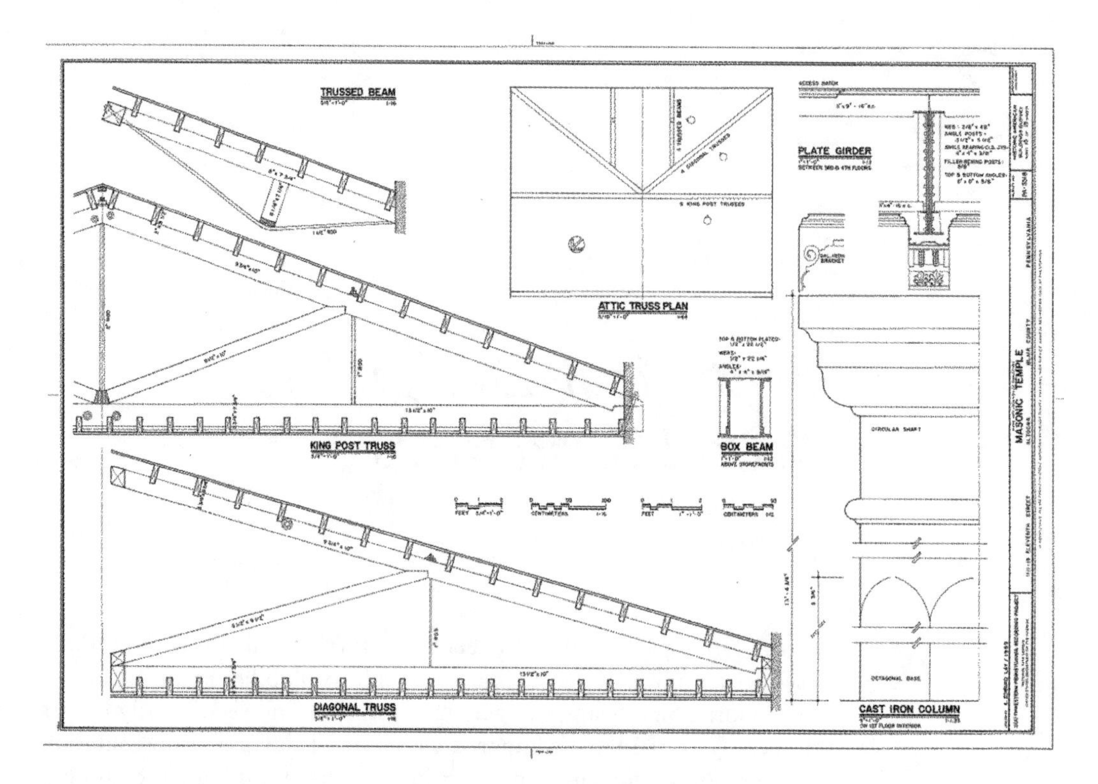

Figure 3-1. Masonic Temple, Altoona, PA, constructed 1889 (HABS PA,7-ALTO,109-).
Source: Historic American Building Survey.

Residential Floors

The sizes for structural elements in residential floors, according to the available textbooks, were often determined on the basis of semiempirical formulas. Hatfield (1871, p. 222) presents a formula for the safe load (in pounds) uniformly dispersed over a simply supported beam:

$$Safe\ load = \frac{2 \times breadth \times depth\ squared \times S \times a}{span\ in\ feet}$$

S in Hatfield's notation is a constant dependent on the breaking strength of the timber (e.g., oak, 315; white pine, 240) used in construction and *a* is the reciprocal of the factor of safety to be used. This formula can be compared to that found in Frank Kidder's book (1886, p. 309):

$$Safe\ load = \frac{2 \times breadth \times depth\ squared \times A}{span\ in\ feet}$$

Carriage-beams, Headers, and Tail-beams.—Fig. 2 represents the plan of the timbers of a floor, having a stairway opening on each side. The short beams, as *KL*, are called the "tail-beams:" the beams *EF* and *GH*, which support the tail-beams, are called the "headers:" and the beams *AB* and *CD*, the "carriage-beams," or "trimmers."

The *tail-beams* are calculated in the same way as ordinary floor-joist; but it is evident that the headers and trimmers will require separate computations.

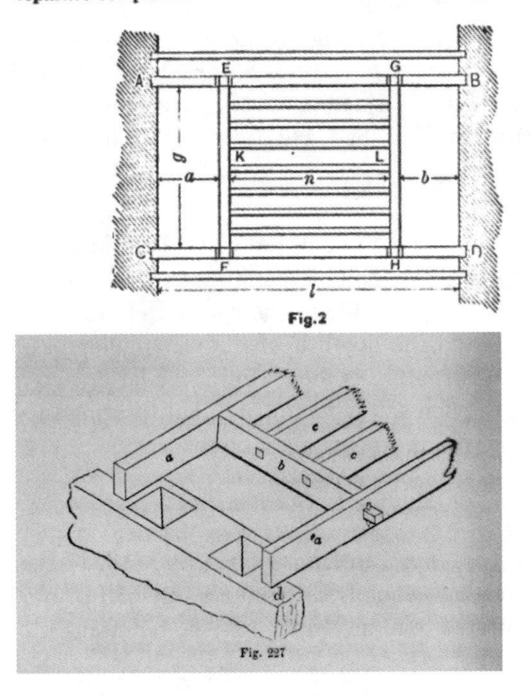

Figure 3-2. Top: Kidder (1886), showing terminology for wood elements used in trimming a stair opening. Bottom: Hatfield (1895), Figure 227, showing a framed opening for a chimney. Note that headers are mortised into trimmers.

This result can be applied to the determination of the required depth for a given load:

$$Square\ of\ depth = \frac{dist\ bet.\ centers \times length\ squared \times (f + f')}{2 \times breadth \times A}$$

where f = live or superimposed load and f' = dead load.

In Kidder's discussion, the values of A are generally one-fourth or less, compared with the values of S used by Hatfield, (e.g., white oak, $S = 574$, $A = 105$; Georgia pine, $S = 510$, $A = 125$). S and A are defined as, respectively, the breaking strength and the allowable stress of a 1-in., 12-in.-long billet loaded at the center. This is numerically 1/18 of the modulus of rupture or safe bending stress in pounds per square inch. Kidder generally recommends the use of safety factors of 3 for dwellings, public buildings, and stores, and further recommends increasing this value by 5/4 for factories and other buildings with long spans.

In *The American House-Carpenter*, Hatfield (1871) gives a large variety of formulas for the size of a column, the width or depth of a beam, and other important design values. Although these formulas are based on notions of the calculation of stresses, they are not directly recognizable as rationally based and in application function much like rules of thumb. The rules are elaborated in his later book, *Theory of Transverse Strains* (Hatfield 1877). Both books and both sets of rules appear to be attempts to transform the designing of beams into a large set of rules of thumb.

An example of another rule for the design of joists is a beam rule, modified and specialized to floors in Hatfield's *Theory of Transverse Strains*. See article 115, p. 89, "Rule for Floors of Dwellings." In this article, Hatfield proposes a total load of 90 lbs/ft^2, composed of 70 lbs/ft^2 live load and 20 lbs/ft^2 dead load. He derives the following rule:

$$180cl^2 = Bbd^2$$

B is the breaking weight of a 1 in^2 piece of wood 1 ft long loaded in the center and is equivalent to 1/18 × breaking stress in pounds per square inch. Example values of B for wood and iron materials are given in a table in the back of Hatfield's book (e.g., Georgia pine, $B = 850$). The breadth b and the depth d of the rectangular beam are expressed in inches, c is the spacing of the joists, and l is their length, the latter two in feet. In the form as shown, the formula has an implicit safety factor of 4. Substituting the allowable bending stress $18F_b$ for B, the formula gives a more modern equivalent:

$$(90 \text{ lb/ft}^2\ c)L^2/8 = F_b bd^2/6$$

In the seventh edition of *The American House-Carpenter* (1871), the same ideas appear in Rule XXXIX: When the weight is equally distributed, one-half of the quotient obtained by the preceding rule (concentrated load at center) may be represented by K:

$$\frac{wl}{2Sa} = K$$

where w is the total of the uniformly distributed load; l is the span in feet; S is a material constant, modified by geometric constants, identical to B previously; a is the inverse of the factor of safety; and K is the required value of bd^2.

In the eighth edition of *The American House-Carpenter* (Hatfield 1895), posthumously edited by the son of the original author, the previous example can be compared with Eq. (28) (p. 109), also stated as Rule XXIII: "Multiply the given weight per superficial foot by the factor of safety, by the distance between the centres of the beams in feet, and by the square of the length in feet; divide the product by twice the value of B for the material of the beams, and the quotient will be equal to the breadth into the square of the depth." Throughout all of Hatfield's work, he derives condition-specific formulas for different loading and support conditions using mixed units and leaving the factor of safety to the designer. He then restates the formulas as rules for application. Although not strictly empirical, in the sense that mechanics is used to derive the formulas, the rules are intended to be used in conventional situations without necessarily having knowledge of the underlying mechanics.

Another carpentry manual, William Allen Sylvester's *Modern Carpentry and Building* (1896), simply states the size of floor joists required for a single-family house. The presumed span is up to 15 ft as illustrated in the plans for Sylvester's own house given in an appendix.

His guide specifications for a residence (p. 182) require 2 × 8 floor timbers, spaced 16 in. apart, and 2 × 6 for attic floors. In his guide for estimating (p. 104), a medium house has floor joists 2 × 8 through 2 × 10, whereas a heavy framed house has 2 × 12 joists for the first floor and 2 × 10 for the second floor. The universal use (even to the present day) of 2 × 4 studs in bearing walls is certainly an empirical procedure. Sylvester prescribes this size for the studs in a bearing wall.

Sylvester also has a brief presentation of design of floor beams by rule, similar to Hatfield's treatment of the topic. For example,

> *If the Dimensions are required to support a Given Weight.*—RULE. Divide the product of the weight and the length in feet by 4 times the safe-load given in the table; the result is the square of the depth multiplied by the breadth or thickness: so we divide this result by the breadth, and extract the square root, which gives the depth. (p. 154)

Pillars

Rondelet's rule for timber pillars (quoted in Stoney 1873) is, "Taking the force which would crush a cube as unity, the force requisite to break a timber pillar with fixed ends whose height is

12 times the thickness, will be	5/6
24 " " " " "	1/2
36 " " " " "	1/3
48 " " " " "	1/6
60 " " " " "	1/12
72 " " " " "	1/24" (p. 442)

Bindon Blood Stoney further cites the rules of R. P. Brereton as "the most useful rule yet published for the strength of large pillars of soft foreign timber with their ends adjusted in the ordinary manner, that is without any special precautions." (p. 443) Brereton's table for fir or pine is shown in Table 3-1, along with a modernization of the units.

The Chicago building code (Chicago City Council 1905, p. 250) specifies wood posts primarily by their *L/d* ratio, with reductions in allowable stress for greater slenderness. Table 3-1 presents a set of similar values from the 1905 version of the Chicago building code (section 1006, "Posts with Flat Ends") and a comparison to the current edition of the additional National Design Specification (NDS) (American Wood Council 2006) for No. 2 pine. The shading represents the actual values reported by Brereton; other values were interpolated. The main difference between the empirical values and the NDS is the sharp falloff in allowable compressive stress at higher *L/d* ratios. This is primarily due to the NDS assumption of a fully pinned end in contrast to the assumption, fundamentally consistent with nineteenth-century experimentation and practice, that the squared ends of large timbers provide rotational restraint. This is partly demonstrated by the results of the square-ended tests by Brereton reported by John Crehore (1886).

The values shown from Kidder (1886) are based on a form of Lewis D. B. Gordon's formula (described in Chapter 9) for rectangular pillars and struts, where the breaking load is a strength value for the species of timber used, divided by the factor $[1 + .004\,(h/d)^2]$. A.

TABLE 3-1. *Various Prescribed and Tested Values of Wood Column Strength*

Ratio of length to least breadth	10	15	20	25	30	35	40	45	50
Breaking weight in tons per square foot of section (Brereton, with interpolations by Stoney 1873)	120	118	115	100	90	84	80	77	75
Brereton's experimental values: square end; factor of safety (FS) = 4, psi	467		447		350		311		
Brereton's results transformed to psi/usual FS = 4	416	409	400	350	312	291	278	267	260
Kidder (1886, p. 222), hard-pine breaking load divided by FS = 4, psi	658	438	481	357	272	212	169	138	113
Chicago building code: white pine and spruce, psi	625					475			300
C_p NDS S-P-F no. 1	0.96	0.90	0.80	0.66	0.52	0.41	0.33	0.27	0.22
NDS Value S-P-F no. 1, psi	601	564	500	414	328	258	205	166	136

Jay Du Bois (1887, p. 375) proposes values of 550 for flat ends and 275 for timber pillars for pinned ends in Gordon's formula, which is equivalent to a valve of .0036 to .0018 for the coefficient of the slenderness ratio in the version of Gordon's formula presented here.

On the basis of the several authors quoted earlier, it is clear that the most important parameter in the design of wood pillars was the ratio of height to width. Although stresses were calculated for various situations, the initial concern in the design of columns was to keep this ratio within acceptable limits, generally less than 20 and in no case greater than 30. The influence of squared ends was also recognized as a factor contributing to the strength of wood pillars. Figure 3-3 shows a nineteenth-century mill building in which apparent pains were taken to ensure that the girders bear fully on the squared ends of the column. The slenderness ratio of the column is most likely large enough that this treatment is needed to ensure the strength of the column. Shown in Figure 3-4 is a measured drawing of an industrial building in which the unsupported length of both the columns and the beams was limited to approximately 15 times the depth of the member.

Trusses

Sylvester (1896) relates truss type directly to span without specifying sizes of timber. He says of a simple king-post truss:

> Figure 51 represents a truss suitable for a span of 30 to 40 feet. The figures indicate stocky elements with a depth approximately 1/16 of the overall truss span. The top chords are

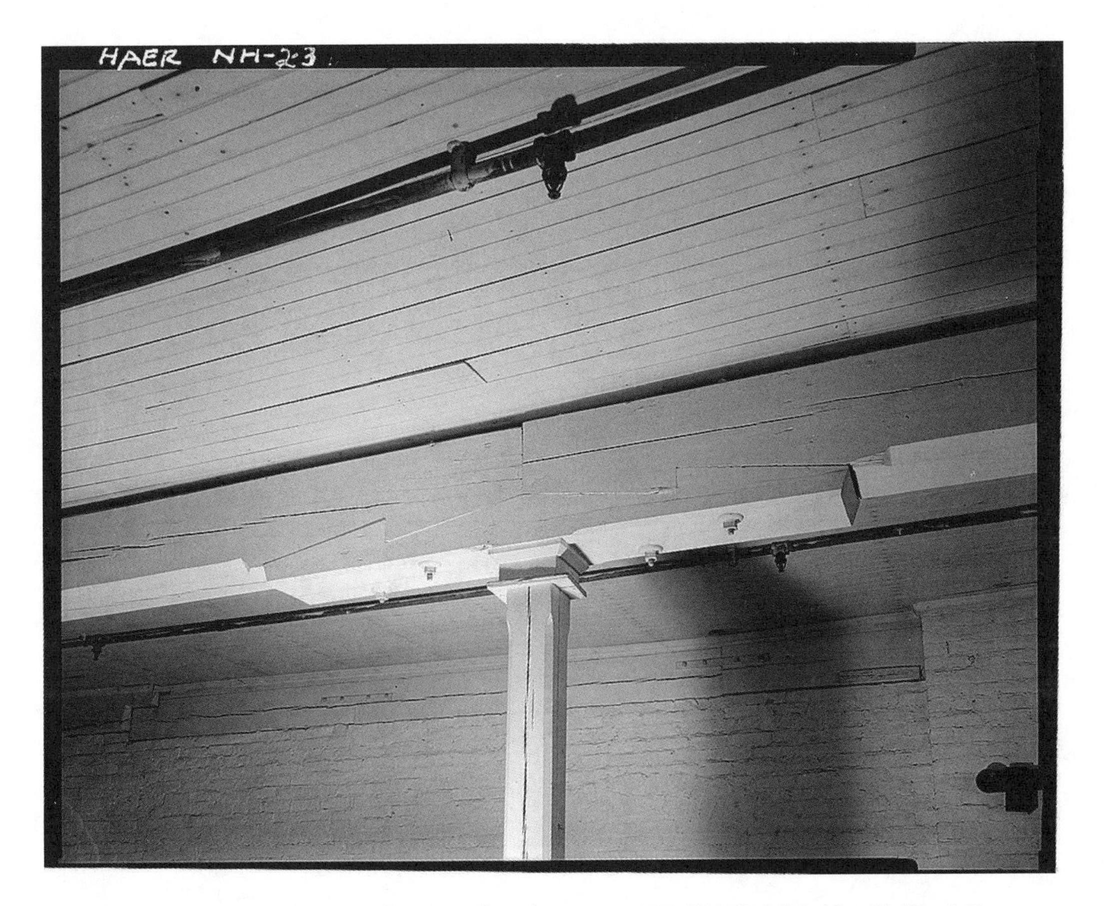

Figure 3-3. Detailed view, Monadnock Mills, Claremont, NH (HAER NH, 10-, CLAR, 6-3).
Source: Photograph by Jet Lowe.

> let into the bottom chord, while the kingpost is rendered as an iron tension member. Other trusses with a greater number of panels, up to a six panel Howe truss, are indicated for larger spaces, such as great halls, or for bridges. (p. 68)

Hatfield (1871) considers the design of several relatively complex trusses (such as those shown in Figure 3-5) by rule, first stating a rule for the determination of the force in a rafter (p. 273); this rule also can be used for the calculation of the top chord force:

> Rule LIII. To obtain the dimensions of the rafter [m]ultiply the value of *R* [vertical strain per foot of surface supported—i.e., vertical component of load] by the span of the roof, by the length of the rafter, and by the distance apart from centres at which the roof trusses are placed, all in feet, and divide the product by the sum of twice the height of the roof multiplied by the value of *P* [transverse compression allowable stress (FS = 4)], set opposite the kind of wood used in the tie beam, added to the difference of the values of *C* [longitudinal compression allowable stress (FS = 4) and *P* in the said table multiplied by 1¼

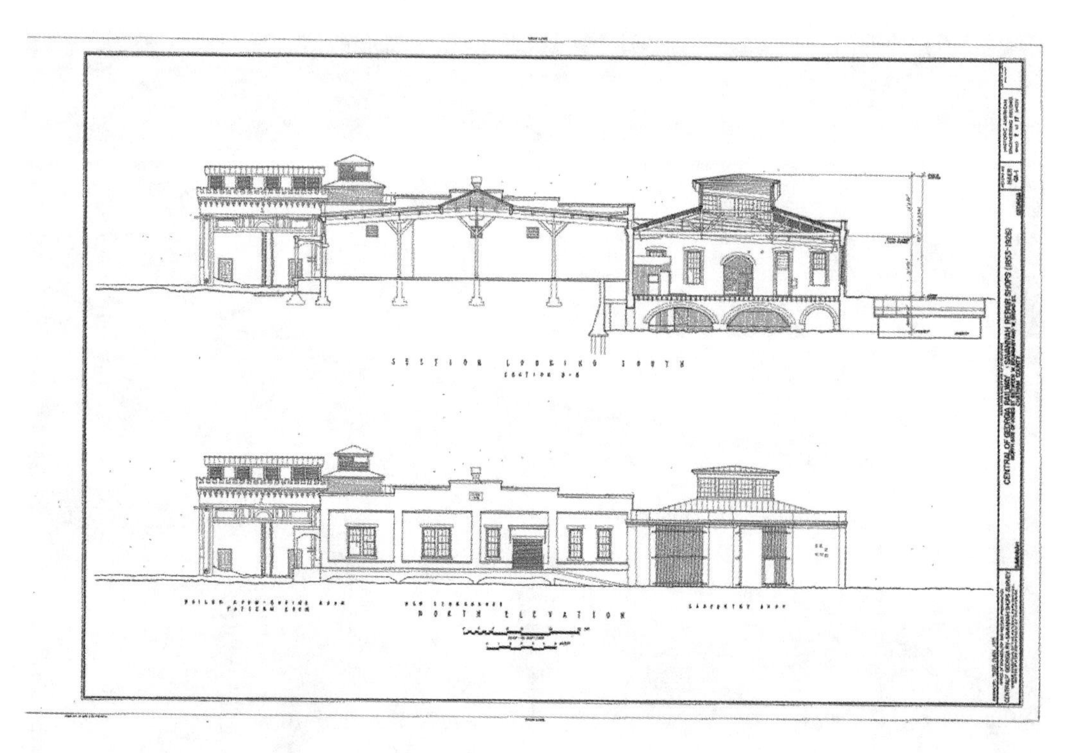

Figure 3-4. Section through Georgia Central of Georgia Railroad repair shops (circa 1853; HAER GA,26-SAV,55-).
Source: Historic American Engineering Record.

> times the length of the arc that measures the acute angle formed between the rafter and a vertical line, the arc having the height of the roof for radius, and the quotient will be the area of the abutting surface of the joint at the foot of the rafter. To the abutting surface add its half, and the sum will be the area of the cross section of the rafter.

The numerator in this complicated expression is double the moment of the loads; divided by twice the height of the roof, this gives the horizontal component of force in the rafter. Beyond this, Hatfield appears to be considering the interaction between transverse and longitudinal compression in sizing the required abutting surface and determining the size of the rafter on this basis. Similarly complicated rules are presented for the determination of the size of the ties and the size of the braces. All of the rules are based on the ad hoc determination of the forces, rather than a systematic treatment of the determination of the forces in all the bars in a truss. For instance, brace force is determined by multiplying the tributary area of the brace times the roof load and adding the force on the adjacent hanger rod. A similar treatment is also given to the attic truss and other truss forms (figure, p. 283).

In Frank Kidder's essay on wooden trusses (1886, p. 392), simple trusses consisting of rafters and ties are to be used for spans up to 24 ft. Beyond this, one may use a braced truss of four panels, or a king-post truss from 25 to 35 ft. The queen-post truss is to be used for

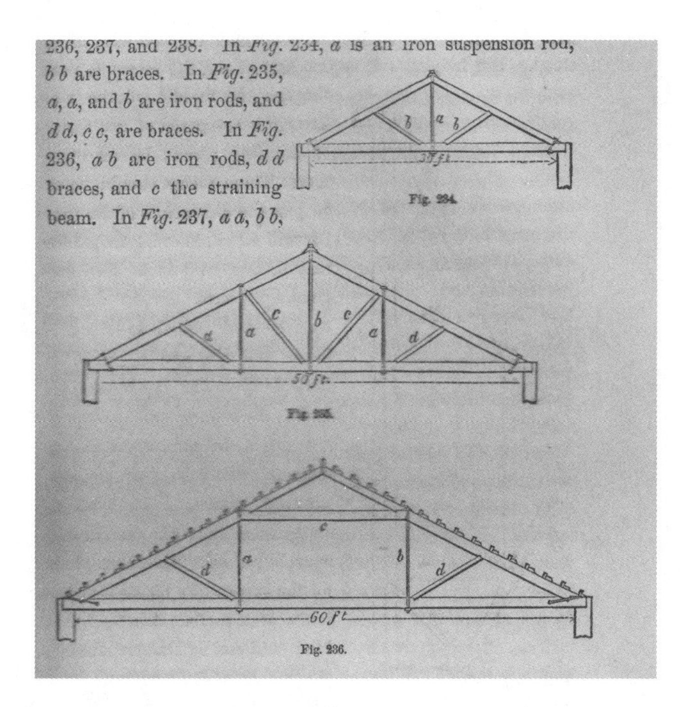
236, 237, and 238. In *Fig.* 234, *a* is an iron suspension rod, *b b* are braces. In *Fig.* 235, *a*, *a*, and *b* are iron rods, and *d d*, *c c*, are braces. In *Fig.* 236, *a b* are iron rods, *d d* braces, and *c* the straining beam. In *Fig.* 237, *a a*, *b b*,

Figure 3-5. Typical trusses.
Source: Hatfield (1895).

spans up to 40 ft, with a modified form of this truss used for spans from 40 to 50 ft. Span-to-depth ratios of 8 or less are recommended for various forms of truss.

Figure 3-6 illustrates a series of trusses designed by a builder-architect, Samuel Bartel, for the reconstruction of a commercial roof (Bartel 1914). The illustration shows empirically designed queen-post trusses in wood. The truss has a span/depth ratio of 6. The top and bottom chords have the same size, although the top chord receives much greater bending stresses. The rods appear to be much larger than necessary at 1 in. in diameter, and the compression members in the web are arguably undersized compared with the top and bottom chords. However, the truss is clearly competent to resist the loads imposed on it for a span of approximately 30 ft.

Empirical design was a fundamental means of designing wood structures in the late nineteenth century, especially relatively simple structures, such as residential framing or floor and roof joists. The empirical design of a wood beam could consist of following a rule, as noted in Kidder, Hatfield, Allen, and other authors, or by simply sizing a wood element in accordance with its span, using a span/depth ratio of around 12. Possibly a greater span/depth ratio can be used for a joist, possibly less for a girder. A pillar certainly can be designed empirically, maintaining a least lateral dimension of about 1/20 the height of the pillar. Even a truss can be designed empirically by selecting the number of panels required on the basis of the span of the truss, maintaining a span/depth ratio of 8, and sizing the top chord on the basis of a beam rule. A significant amount of design of wood structures was accomplished on this basis.

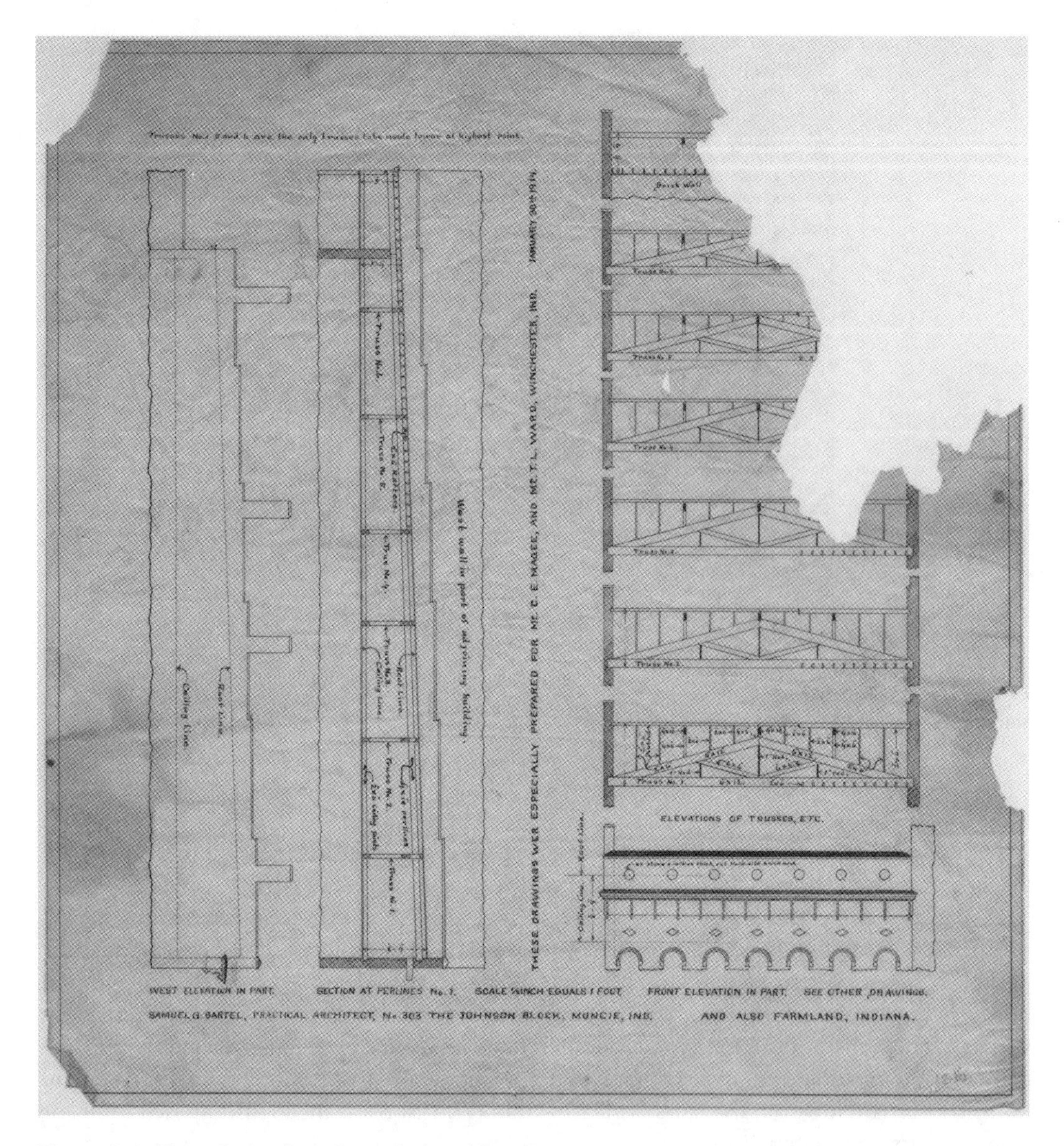

Figure 3-6. Truss design by a "practical architect."
Source: Bartel (1914). Reproduced by permission of the Ball State University Libraries.

References Cited

American Wood Council. (2006). *National design specification for wood construction.* American Forest and Paper Association, Washington, DC.

Bartel, S. G. (1914). "New roof for building elevations and section." Drawings and Documents Archive, Samuel G. Bartel Collection, Ball State University, Muncie, IN.

Chicago City Council. (1905). *An ordinance relating to the department of building.* Moorman and Geller, Chicago.

Crehore, J. D. (1886). *Mechanics of the girder.* John Wiley and Sons, New York.
Du Bois, A. J. (1887). *The strains in framed structures*, 4th Ed. John Wiley and Sons, New York.
Hatfield, R. G. (1871). *The American house-carpenter*, 7th Ed. John Wiley and Sons, New York.
Hatfield, R. G. (1877). *Theory of transverse strains.* John Wiley and Sons, New York.
Hatfield, R. G. (1895). *The American house-carpenter,* 8th Ed. John Wiley and Sons, New York.
Kidder, F. (1886). *The architects' and builders' pocket-book*, 3rd Ed. John Wiley and Sons, New York.
Stoney, B. B. (1873). *The theory of strains in girders and similar structures.* Van Nostrand, New York.
Sylvester, W. A. (1896). *Modern carpentry and building.* Allen Sylvester, Boston.

4

Empirical Design of Iron and Steel Structures

Because iron and steel structures usually were designed, fabricated, and constructed as simply supported beams and columns, the design of such structures relied on analysis to a greater extent than did the design of masonry and timber structures. Nevertheless, empirical design played an important role in the design of iron structures, both in selecting initial proportions and in the application of rules to the final proportioning of structural elements. Empirical rules for metal structures took several forms. The first form was the simply prescriptive: Isami Hiroi's rule for the depth of a girder, given following, is an example of this approach. A widespread practice also existed of judging metal structures, especially beams and girders, by span/depth ratios, whereas the actual design may have been accomplished by one of the analytical procedures outlined in Chapter 8. Finally, there was widespread development of rules of thumb, similar to those used for wood structures (Chapter 3), whereby a more or less rational procedure is reduced to a rule for design written in variables that express load, dimensions of the area supported, dimensions of the beam or column, and strength of the material. Hiroi's rule for the depth of a girder certainly also contains elements of this approach.

Various authors advance similar rules for the appropriate span/depth ratios for girders. Bindon B. Stoney (1873, p. 322) gives the normal span/depth ratio of a plate girder as about 15. William Humber (1869, p. 25) says that plate girders vary in depth from 1/10 to 1/16 of the span and that the most economical depth is generally 1/12 of the span, whereas continuous girders vary in depth from 1/15 to 1/20 of the span. Stoney (article 461, p. 349) specifies that the depth of girders varies from 1/8 to 1/16 of the span, with 1/10 to 1/12 being the most common. He further suggests distributing three-fourths of the area of the girder cross section to the web on the grounds of its relative contribution to strength. According to this notion, a double web box section with ¼-in. web plates 24 in. deep should have flange plates ¼ in. × 8 in. It is more efficient to add depth to the web than to add material to the flange and to keep the depth of the girder the same. F. W. Sheilds (1867, p. 44) gives the maximum economical length for a plate girder as opposed to a lattice girder as 90 ft. Span to depth ratios, according to Sheilds, should range from 10 : 1 to 15 : 1 for plate girders. Kidder (1886, p. 285) says that the deepest beam is always the most economical. Hiroi (1893, p. 46) gives the range of plate girder depths as 1/9 to 1/12 of the span, with the former used for shorter and the latter for longer spans. According to Hiroi, a plate girder is uneconomical beyond a span of about 100 ft.

Sizes of the elements of girders also were dictated by the availability of stock. Stoney (1873, p. 322) argues in favor of plate girders for simple spans over trussed girders; he says further that economy of plate girders is better in shallow rather than in deep girders. The recommended minimum thickness of a plate web is ¼ in. Stoney (p. 326) recommends height/thickness of vertical plates of 15 unless provided with stiffeners. In cast-iron girders, the web is relatively thicker and contributes relatively more to the overall strength of the girder. There are other limitations on the size of elements that can be fabricated into girders, which limits the range in which a filled web or plate girder can be applied. Stoney (article 439, p. 326) discusses maximum bar and plate sizes of iron available from rolling mills. If, according to Stoney, a builder has to pay a premium for plate iron more than 4 ft wide (in addition to the extra thickness required in a girder web), then it is likely that 4 ft times the span/depth ratio of 12 : 15 gives a practical limitation of 50 to 60 ft on the span of a plate girder. The span capacity of a trussed girder was greater: it will be shown in Chapter 7 that a span of 200 ft was routinely attainable with a trussed girder.

Mansfield Merriman and Henry S. Jacoby (1894) give an economical depth for a plate girder as the depth at which the weight of the web is approximately equal to the weight of the flanges. Kidder (1886, p. 346) calculates strength of riveted girders by an empirical rule, similar to those presented for wood girders in the previous chapter:

$$\text{Safe load in tons} = 10 \times \text{area of one flange} \times \text{height}/(3 \times \text{span in feet})$$

The same formula, inverted, gives

$$\text{Area of one flange in square inches} = 3 \times \text{load} \times \text{span in feet}\,/(10 \times \text{height of web in inches})$$

The web plate is similarly designed: divide total uniformly distributed load, in tons, by the vertical sectional area of the web plate. If this number exceeds the number in a table, then stiffeners are required up to within one-eighth span from the middle of the girder. The entries in Table 4-1 represent average shear stress in the web, compared with the maximum permissible height/thickness ratio for the web. Kidder explains the information in Table 4-1 as follows:

> The height of the girder is measured in inches and is the height of the web-plate, or the distance between the flange-plates. The web we may make either one-half or three-eighths of an inch thick; and, if the girder is loaded with a concentrated load at the centre or any

TABLE 4-1. *From Kidder (1886, p. 347) for Determination of Stiffener Requirement in Plate Girders*

$l \times 1.4$	30	35	40	45	50	55	60	65	70	75	80	85	90	95	100
t	3.08	2.84	2.61	2.39	2.18	1.99	1.82	1.66	1.52	1.40	1.28	1.17	1.08	1.00	0.92

other point, we should use vertical stiffeners the whole length of the girder, spaced the height of the girder apart. If the load is distributed, divide one-fourth of the whole load on the girder, in tons, by the vertical sectional area of the web-plate; and if the quotient thus obtained exceeds the figure given in the following table, under the number nearest that which would be obtained by the following expression (1.4 × height of girder/thickness of web), then stiffening pieces will be required up to within one-eighth of the span from the middle of the girder (p. 347).

Kidder's example is as follows:

A brick wall 20 feet in length and weighing 40 tons, is to be supported by a riveted plate-girder with one web.

$$\text{Flange area} = (3 \times 40 \times 20)/(10 \times 24) = 10 \text{ square inches}$$

Subtracting 5 square inches for 2 3 × 3 [× 3/8″] angle irons, we have 5 square inches: use 5/8″ × 8″ flanges

Use 3/8″ web and put two stiffeners at the ends of the girder

To find if it will be necessary to use more stiffeners, divide 1/4 of 40 tons by the area of the vertical section of the web, 9 square inches, and obtain 1.11.

The expression 1.4 × height of girder/thickness of web = 89.6, say 90, and the figure under it is 1.08, which is less than 1.11, showing we must use vertical stiffeners up to within 3 feet (1/8 span) of the center of the girder. (Kidder 1886, p. 347)

Kidder (p. 350) also describes the construction of cast-iron/wrought-iron, hybrid tied-arch girders for the support of masonry. Construction considerations for this type of fitting are particularly important. The tie rod is pretensioned by heating prior to anchoring. These rods are proportioned by allowing 1 in.^{2} of cross section for every 10 net tons of load imposed on the span of the arch, based on a span/rise ratio of 10.

As a further example of the use of empirical procedures in the design of iron structures, Figure 4-1 illustrates Louis Sullivan's preliminary design for the Farmer's and Merchant's Bank in Columbus, WI. The transverse section shows that the architect chose and a 12-in. (31.5 lbs/ft) I shape that is compatible with the 28-ft transverse span of the building.

Isami Hiroi (1893) gives Octave Chanute's rules for girder webs: web thickness 1/80 the height stiffener spacing no more than two times the web height. Hiroi also presents strictly empirical rules for the estimation of the weight of a plate girder, at least between 20- and 80-ft span (which might be taken as limits on the span of a filled plate girder at the time):

single-track deck girder

$$W = 10\,(s^2 + a)$$

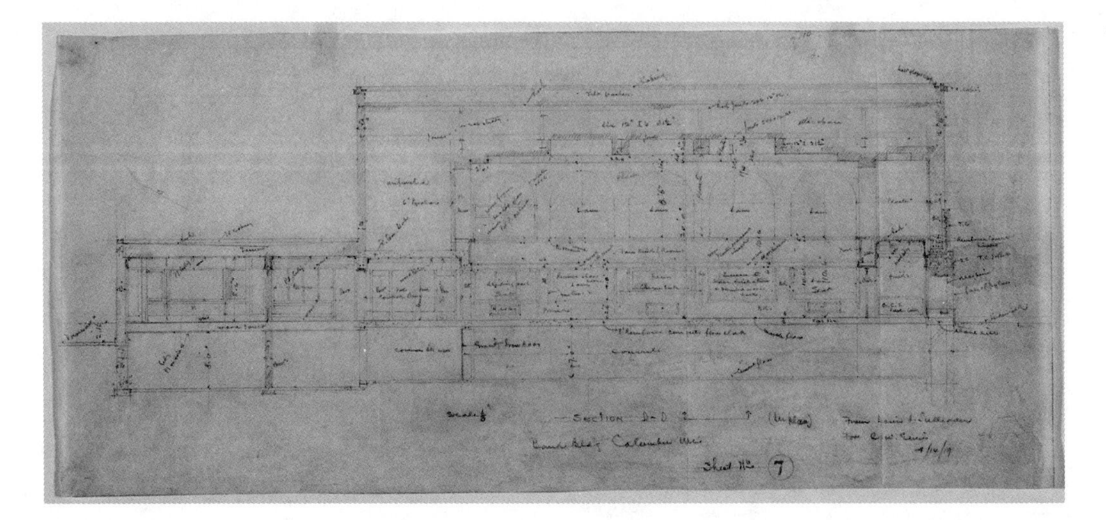

Figure 4-1. Sullivan's preliminary design for the Farmers and Merchants Bank, Columbus, WI (1911).
Source: Avery Architectural and Fine Arts Library, Columbia University.

For through girder with floor beams and stringers, this value of W is modified according to

$$W' = W + 300\,s$$

where W is the girder weight in pounds, *s* is the span in feet, and *a* is a constant from 200 to 330 that depends on floor system and loading. (Hiroi 1893, p. 31)

Cast-iron girders and beams were treated separately due to the lesser tensile resistance of the material. Most authors worked with a tensile strength of about one-sixth of the compressive strength and so recommended apportioning six times the area of the top flange to the bottom flange.

Charles Haslett and Charles Hackley (1859, p. 212) give a rule for the determination of the breaking strength of a cast-iron beam proportioned according to this practice. The best dimensions of a cast-iron beam have a bottom flange area of six times the top flange area. The total distributed breaking weight is equal to the area of the bottom flange in inches times the depth of the beam in inches, divided by the span in feet. This can be compared to Fleeming Jenkin's (1876) account of Hodgkinson's rule: $M = 16{,}500S_t d$, for which the implied maximum stress = 8.25 tons/in.^{2} (breaking, or ultimate, stress), and in the units of Haslett and Hackley, the breaking weight in tons is $W = 2.75S_t d/L^*$, L^* in feet. In these formulas S_t is the area of the tension flange in square inches, d is the depth of the girder, and L^* is the span of the beam. Kidder (1886) reiterates a version of Hodgkinson's rule in a rule for the strength of a cast-iron beam:

$$\text{Breaking load in tons} = \text{area of bottom flange (in.}^2) \times \text{depth (in.)} \times 2.166/\text{clear span in feet}$$

Continuous Beams

Properties of continuous beams, primarily intended for use in wood framing but also applicable to metal structures, are presented as simple rules in Kidder (p. 233). One of Kidder's examples follows:

> STRENGTH.—*Continuous girder of* TWO *equal spans, loaded uniformly over each span*
>
> $$\text{Breaking weight} = \frac{2 \times B \times D^2 \times A}{L}$$
>
> A represents a material strength (1/18 allowable tension ÷ 3), which needs to be multiplied by the safety factor of 3 if the breaking weight is the quantity sought. B and D are width and depth of the beam (in inches), and L is a single span in feet.
>
> The coefficient A is referred to the strength of a beam of one-foot span, subjected to a mid-span load. The moment of resistance is Abd^2, while the uniform load that can be supported is twice the concentrated load.
>
> The remainder of the breaking loads are given similarly in proportional form:
>
> Two equal spans, concentrated load at center, 4/3
> Two equal spans, uniformly distributed load at center, 5/2
> Three equal spans, concentrated load at the center, 5/3

Columns

The formulas developed by Hodgkinson (1846), William John Macquorn Rankine (1877, p. 360), and Lewis D. B. Gordon were used both for the design of iron columns and for the design of wood columns. The most widely used formula for the design of columns is very generally attributed to Gordon (1815–1876). The application of Gordon's formula is discussed in detail in Chapter 9. As in the design of wood columns, the designer must be aware of the use of pinned end values versus fixed, or partially fixed values, where columns with flat ends are considered. Gordon's formula also uses empirically developed coefficients for the modification of l/h ratios for various configurations of columns. Figure 4-2 shows an illustration from A. Jay Du Bois (1887) of the modifications applied to Gordon's formula for cross-section properties and end conditions. The modification for open-latticed channel struts with flat ends is the substitution of 4,880 in the denominator and 2,440 for pinned ends. The corresponding values for single I bars are 1,720 and 860. Du Bois credits these values of the breaking strength of wrought iron columns to Shaler Smith and notes that a further safety factor of $6 + l/20d$ should be applied.

John Davenport Crehore (1886) refers to Hodgkinson's and Gordon's formulas for the strength of wrought iron and cast-iron girders as "empirical." Although the application of these formulas certainly has empirical elements, particularly in the development of the coefficients in Figure 4-2, the main discussion of the Rankine–Gordon rule for column strength is reserved for Chapter 9.

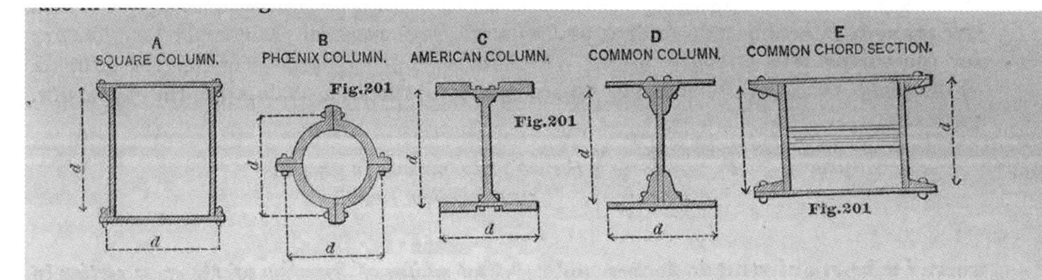

For these forms, the following special formulæ are recommended by C. Shaler Smith for *wrought iron;* where d is the least dimension of the rectangle enclosing the cross section, and l is the length, both in inches.

	A.	*B.*	*C.*	*D.*	*E.*
Flat ends,	$\frac{38500}{1+\frac{l^2}{5820\,d^2}}$,	$\frac{42500}{1+\frac{l^2}{4500\,d^2}}$,	$\frac{36500}{1+\frac{l^2}{3750\,d^2}}$,	$\frac{36500}{1+\frac{l^2}{2700\,d^2}}$,	The pin being so placed that the moment of inertia is, as near as practicable, equal on both sides of same, use formula for square column.
One pin end,	$\frac{38500}{1+\frac{l^2}{3000\,d^2}}$,	$\frac{40000}{1+\frac{l^2}{2250\,d^2}}$,	$\frac{36500}{1+\frac{l^2}{2250\,d^2}}$,	$\frac{36500}{1+\frac{l^2}{1500\,d^2}}$,	
Two pin end,	$\frac{37500}{1+\frac{l^2}{1900\,d^2}}$,	$\frac{36600}{1+\frac{l^2}{1500\,d^2}}$,	$\frac{36500}{1+\frac{l^2}{1750\,d^2}}$,	$\frac{36500}{1+\frac{l^2}{1200\,d^2}}$,	

The safe working stress is found by dividing the "crippling stress," as determined by the above formula, by $4 + \frac{l}{20\,d}$, where l is length in inches, and d is least dimension of enclosing rectangle.

Figure 4-2. Empirical modifications to Gordon's formula for columns of various shapes.
Source: Du Bois (1887).

There are other elements of empirical design in the sizing of steel columns and compression members. The notes of the engineers of the Berlin Iron Bridge Company in East Berlin, CT, indicate that they did not reduce top chord allowable compressive stresses for increased length of these elements. Based on examination of the notes that are available, they used full allowable stress and used larger box sections to make it unnecessary to apply a reduction formula. The required reduction factor obtained, for instance, from a wrought iron box section 16 in. × 12 in., with a length of 190 in., should be, according to Du Bois's version of the Gordon formula coefficients, 88%. From the records of the company, in the strain sheet for the bridge at North Anson, ME, the calculations indicate a strain of 77.0 tons, and the design calls for a cross-sectional area of 14.8 in^2, resulting in an unreduced stress of 5.17 tons/in^2. The effect is more dramatic on smaller bridges. A five-panel 80-ft span bridge at Suffield, CT, has a span/width ratio of about 30, resulting in a reduction factor of 0.67, yet the top chord stress is 4.76 tons/in.^{2}

It is difficult to find explicit rules for the proportioning of trusses, although it is equally difficult to find iron trusses from this time period with a span/depth ratio greater than 7 or

8. Conversely, as is discussed in Chapter 7, there is a significant body of literature on the optimum proportioning of *panels* within trusses. Both struts and ties are most economical when inclined at 45 degrees (ties) or 50 degrees from the horizontal (struts). The result of this is that panels for single-system trusses are approximately square, and the number of panels for such trusses is effectively limited to 8, resulting in a span limit of approximately 140 ft. To span greater distances, it becomes necessary, as discussed in Chapter 7, to use a double system.

To repeat the discussion at the beginning of this chapter, empirical design is less in evidence for metal structures compared with the practice of designing wood and masonry structures by empirical methods. However, elements of empirical design, especially preliminary design by the use of appropriate proportions, and the application of empirical rules that may have been derived more or less analytically, are well established in the late nineteenth-century structural design profession.

References Cited

Crehore, J. D. (1886). *Mechanics of the girder.* John Wiley and Sons, New York.

Du Bois, A. J. (1887). *The strains in framed structures*, 4th Ed. John Wiley and Sons, New York.

Haslett, C., and Hackley, C. (1859). *The mechanic's, machinist's, and engineer's practical book of reference.* W. A. Townsend, New York.

Hiroi, I. (1893). *Plate-girder construction*, 4th Ed. Van Nostrand, New York.

Hodgkinson, E. (1846). *Experimental researches on the strength and other properties of cast iron.* John Weale, London.

Humber, W. (1869). *A handy book for the calculation of strains in girders.* Van Nostrand, New York.

Jenkin, F. (1876). *Bridges: An elementary treatise on their construction and history.* Adam and Charles Black, Edinburgh.

Kidder, F. (1886). *The architects' and builders' pocket-book*, 3rd Ed. John Wiley and Sons, New York.

Merriman, M., and Jacoby, H. S. (1894). *A treatise on roofs and bridges*, part 3, 1st Ed. John Wiley and Sons, New York.

Rankine, W. J. M. (1877). *A manual of applied mechanics,* 9th Ed. Charles Griffin, London.

Sheilds, F. W. (1867). *The strains on structures of ironwork*, 2nd Ed. John Weale, London.

Stoney, B. B. (1873). *The theory of strains in girders and similar structures.* Van Nostrand, New York.

Part II

Analytical Methods

5

Introduction to Analytical Computations in Nineteenth-Century Engineering

Analytical Methods

The introduction, use, and choice of analytical methods in the nineteenth century were subject to the large computational effort required to implement an analytical method. Although the theory of structural analysis was sufficiently well advanced by the later nineteenth century to manage problems, such as beams continuous over supports, arches with fixed supports, statically indeterminate trusses, multiple portal frames, and other such problems that routinely arose in practice, analytical methods for these structures required a very large amount of exacting calculations. Even as widespread and simple a structure as the masonry arch is 3 degrees statically indeterminate, and the curvature of the structure's axis with respect to the loading made difficult the computation of internal forces and bending moments along an arch, masonry or iron. The principal means of accomplishing such calculations was by the development of tables, with specialized individuals (known as computers) in engineering offices required to calculate the entries in the table. Some examples of this type of computation are shown in Chapter 6, which covers the application of analytical methods to the masonry arch. The most prominent iron arch of the time was certainly the Eads Bridge in St. Louis, known as the St. Louis Arch. Several monographs were written by James Buchanan Eads on the computation of the stress in the St. Louis Arch (Eads 1868) as well as by others, for example Malverd A. Howe (1906). A. Jay Du Bois, who is described in Chapter 11 and onward as a proponent of graphical methods, describes this type of large computational effort skeptically: "Those acquainted with the analytical investigation of the 'braced arch' as contained in [Eads 1868] will not, we feel sure, be slow to recognize the advantages of the present [graphical] method." (Du Bois 1877, p. vii)

Trusses, however, were routinely calculated, and methods were developed to write the forces in the members of a truss directly on a diagram to produce a "strain sheet." As a result, truss computations were not as involved as arch computations. A detailed inquiry into truss computations is given in Chapter 7. Analytical methods also existed for the support moments in multispan girders, which were known to be statically indeterminate. The widespread Clapeyron's method, or the three-moment equation, were frequently used for computations of this nature, as described in Chapter 8.

Modern methods of analysis are based on increasingly precise computations, where efficiency is less critical because the digital computer is the primary calculating instrument. Because of the difficulty of computations in the late nineteenth century, however, methods from this period sometimes show a remarkable economy of computational effort. There were two fundamental realms of computation: analytic and graphic. In the analytical realm, engineers of this period had limited resources for computation, primarily the 10-in. slide rule (three-digit precision) for multiplication, division, extraction of roots, and for trigonometry; the 20-in. slide rule (four-digit precision) for similar tasks; various forms of adding machine for addition, multiplication, subtraction, and division; the Thacher calculator (five-digit precision); or published multiplication and reciprocal tables for greater precision in computation.

The illustration in Figure 5-1 shows the use of a 20-in. slide rule for the simple operation of multiplication (453 × 217) and division. The illustration in Figure 5-2 shows how a Thacher calculator can be used for greater precision for the same operation. Successive operations on a Thacher calculator, such as a multiplication followed by a division, required either writing an intermediate result or multiplying by a reciprocal. Published tables of four- or five-digit reciprocals were available for this purpose (Oakes 1865). A. L. Crelle (1897) published tables of the products of all pairs of three-digit numbers, which could be used for rapid multiplication. An example of the multiplication of two three-digit numbers on a Crelle table is shown in Figure 5-3. For larger numbers the procedure involved parsing the multiplicands into three-digit numbers. For instance, to multiply 35,453 and 73,217 requires the computer to multiply 217 × 453, 217 × 35, 73 × 453, and 73 × 35. Each of these four numbers can be looked up quickly and written in the correct alignment to be added by hand or by using an instrument. That is, the computer would write, in the order as given,

```
      98 301
    7 595
   33 069
 2 555
```

and then sum these quantities to obtain the result 2,595,762,301.

A computer placed Crelle's results in tables. Tables often included 10 to 20 columns of calculations, as required for the assembly of a complex equation. Some of the analytical arch calculations shown in Chapter 6 exemplify this type of calculation. To satisfy an equation such as the following example from the analysis of an arch requires all of the

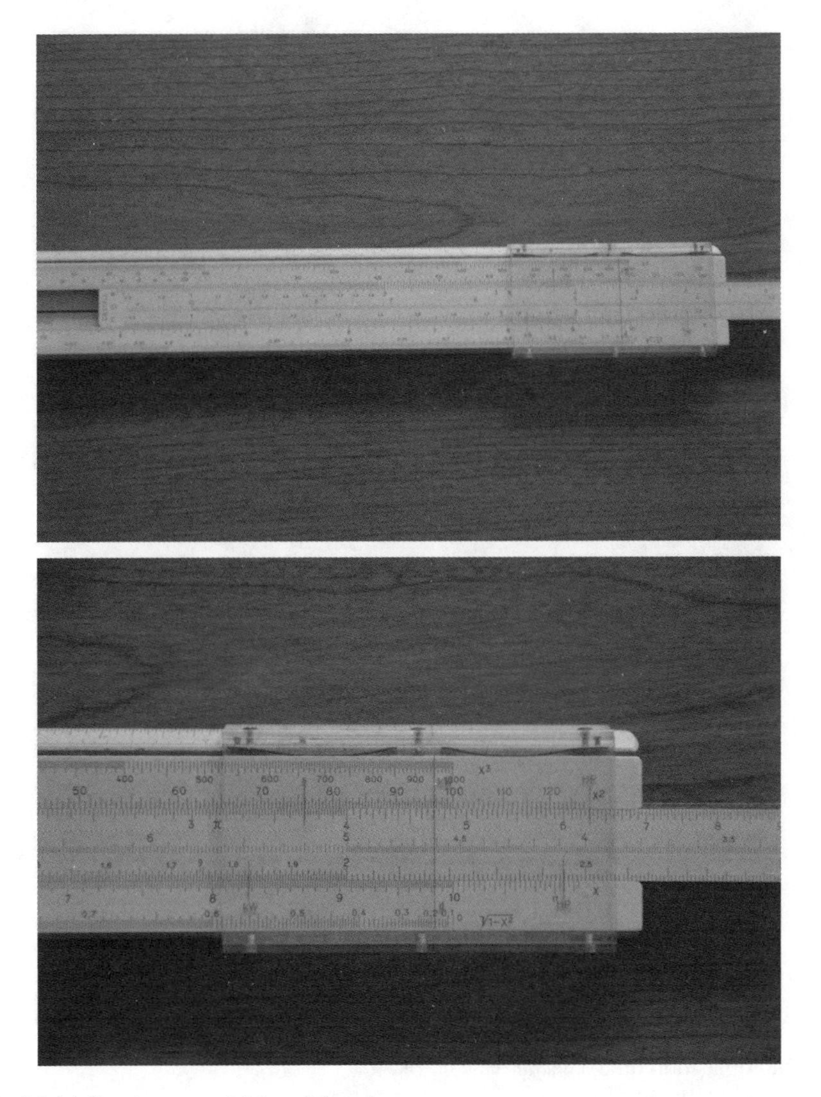

Figure 5-1. Multiplication on a 20-in. slide rule.
Source: Photograph by the author.

variables and their summations to be put into a table such as Table B6-1-4, which has 18 columns:

$$\frac{(R_1 - R_2)\sum_0^a\left(\frac{l}{2} - x\right)x + 2R_2\sum_a^{l/2}\left(\frac{l}{2} - x\right)^2}{\frac{4}{l}\sum_0^{l/2}\left(x - \frac{l}{2}\right)^2}$$

According to the author presenting this large-scale tabular computation, Malverd Howe (1914), most of the table entries are calculated by the method of differences, with

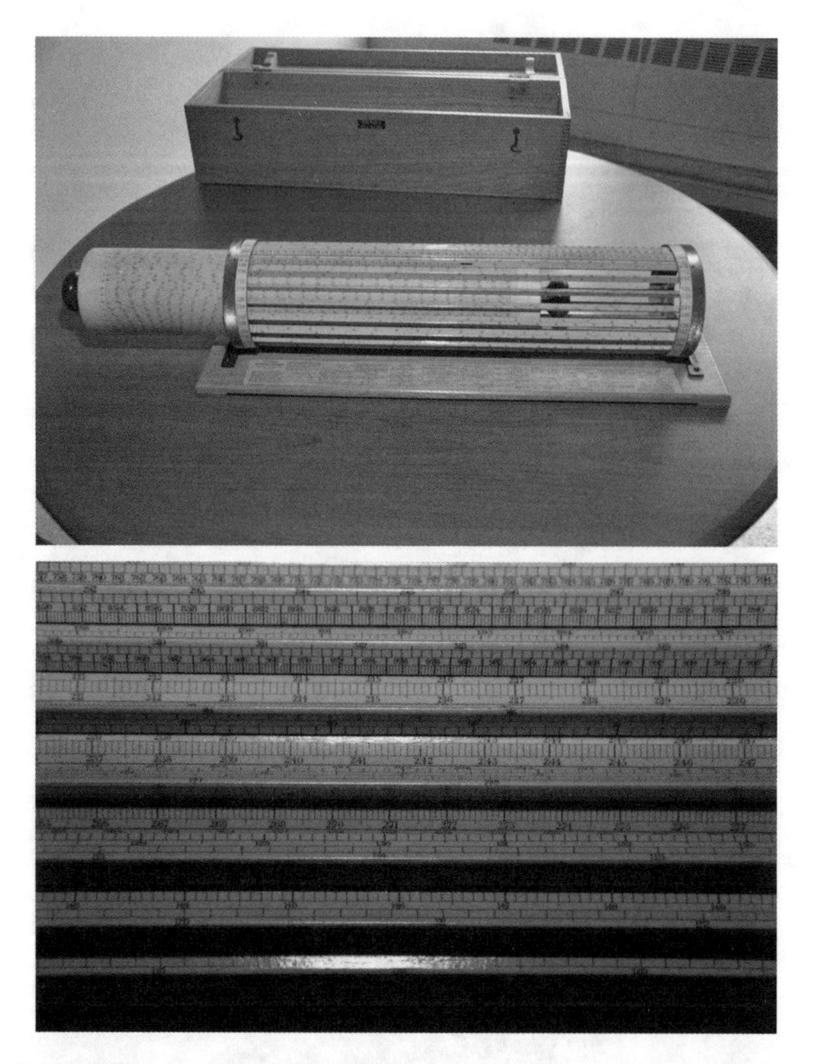

Figure 5-2. Multiplication on a Thacher calculator.
Source: Photograph by Paul Kremer.

occasional complete calculations for the verification of the values. The method of differences involves considering the trend of the values in a table by the methods of differential calculus and estimating subsequent values by extrapolating the truncated series approximation of the solution. An example of a practical application of differential methods to calculations in the nineteenth century is the work by James Pearson (1849).

Analytical methods were considered particularly valuable because of the precision of the answers compared with either of the two other methods available. Empirical design, obviously, gives answers that appear to be imprecise and may generally have been considered uneconomical relative to answers that were developed by other methods. Graphical analysis, although necessary for certain types of structures, gives answers that may have errors that result from imprecise measurement of lines or angles, resulting in a possibility of errors greater than the results of analysis.

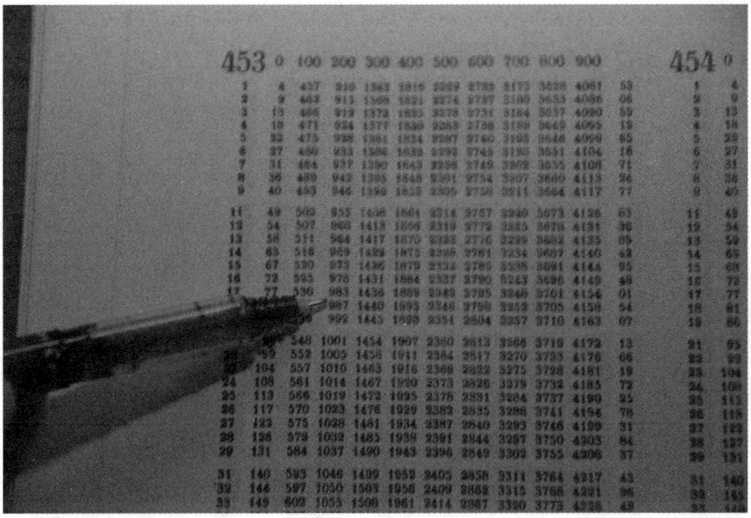

Figure 5-3. Multiplication of three-digit numbers using a Crelle table.
Source: Photograph by the author.

The disadvantage of analytical methods, of course, was the length and tedium of the calculations that accompany this analysis. The example of the analysis of a bridge arch is given in Chapter 6. In Eads's summary of the analysis of one of the arches of his bridge, he presents 20 pages of symbolic computations to determine the forces in the ribbed arch under uniform and concentrated loads and follows this with an additional 10 pages of tabular computations to calculate the desired numerical results (Eads 1868). In the application of the three-moment equation to the analysis of a multispan bridge, described in detail in Chapter 8, it is necessary to calculate a series of coefficients equal to the number of spans and to apply these coefficients sequentially to each of the spans while keeping separate all of the possible loading conditions of dead load and live load in each of the spans. Although a similar graphical solution is available, as described in Chapter 13, it was seldom employed.

As with the other methods explored in this book, analytical methods had a particular place in the toolbox of the late nineteenth-century engineer. These methods were particularly used for truss bridges and were necessary when continuous girders were employed. Analytical methods also were used in the analysis of columns and were an alternative to graphical methods in the analysis of building trusses and the analysis of arches.

References Cited

Crelle, A. L. (1897). *Dr. A. L. Crelle's calculating tables giving the products of every two numbers from one to one thousand and their application to the multiplication and division of all numbers above one thousand*. David Nutt, London.

Du Bois, A. J. (1877). *Elements of graphical statics*, 2nd Ed. John Wiley and Sons, New York.

Eads, J. B. (1868). *Report of the engineer-in-chief of the Illinois and St. Louis Bridge Company*. Missouri Democrat Book and Job Printing House, St. Louis, MO.

Howe, M. A. (1906). *Treatise on arches*, 2nd Ed. John Wiley and Sons, New York.

Howe, M. A. (1914). *Symmetrical masonry arches*. John Wiley and Sons, New York.

Oakes, W. H. (1865). *Table of reciprocals of numbers from 1 to 100,000*. Charles and Edwin Layton, London.

Pearson, J. (1849). *The elements of the calculus of finite differences*. E. Johnson, Cambridge, UK.

6

Analysis of Arches

The determination of the size of critical elements in arches was a frequently recurring problem in the latter part of the nineteenth century. Arches were often used for the construction of permanent bridges and were also widely used in building facades. An example of a significant arch structure from this time period is shown in Figure 6-1. Two analytical methods for arches were prevalent in the late nineteenth century: strictly analytical and semigraphical. The semigraphical methods require some knowledge of graphic analysis, which is presented formally in Chapter 12, so only a brief introduction to this topic is presented at the conclusion of this chapter. The analytical methods are primarily addressed through the methods presented in William John Macquorn Rankine's *Manual of Civil Engineering* (1876a). Although this is an English work, it was widely distributed and widely used in the United States. It was previously encountered in Chapter 2, where the subject was empirical methods. Some work with Rankine's analytical method may have been sufficient to convince an engineer of the merit of using empirical formulas instead of the exact analytical formulas for the analysis of the arch. Further reference will be made to an American work, Malverd Abijah Howe's *Treatise on Arches* (1897). The presentation of the semigraphical methods will refer to Frank Kidder's *Architects' and Builders' Pocket-Book* (1886), in combination with Ira Osborn Baker's *A Treatise on Masonry Construction* (1907), in addition to other works. Kidder (1886) is exclusively devoted to building structures, whereas Baker (1907) is authored by one of the authorities on masonry construction in the United States in the late nineteenth and early twentieth centuries.

Figure 6-1. Example of a railroad masonry arch bridge. Baltimore and Ohio Railroad, Benwood Bridge, Marshall County, WV (HAER WVA, 26-BEN, 1–4).
Source: Photograph by William E. Barrett.

Analytical Methods

Simple, qualitative analytical methods that a modern engineer would recognize as a collapse mechanism analysis were widely known in the analysis of arches, although they seem to be relatively little used—more widely used methods include the strictly empirical methods described in Chapter 2 and the fully analytical methods described in this chapter. Hermann Haupt (1856, p.125), in his early American-oriented treatise on bridge construction, presents observations on the arch on the basis of a collapse analysis, attributed to Emiland M. Gauthey (1771), in which he solves the horizontal thrust of the triangle of opposing forces in the two sides of a symmetric arch about the joints of rupture. The joint of rupture is a point in the intrados (inside face) near the abutment, where the failure is initiated in the part closer to the center of the arch. That part of the arch between the joint of rupture and the abutment is simply considered part of the abutment. The remaining part of the arch between the joints of rupture is considered resisting and is designed for the horizontal thrust calculated. The conclusion to Haupt's analysis is that "the former notion about the arch being perfectly equilibrated by a catenarian curve, is now regarded as a fallacy." (p. 127)

Haupt cannot agree entirely with such a radical statement, saying that, while the catenary should not be used to shape the intrados, it may be used to find the proper direction of the joints.

Rankine (1876b), in his *Manual of Applied Mechanics*, develops a strictly analytical treatment of iron arch ribs, which he subsequently applies to masonry arches. The formulas that he develops are computationally complex and can only be adapted to a few special cases. He does make important inferences regarding his own empirical formula for arch proportions from one of the analytical formulas. Rankine considers both the initial curve of the arch and the variations in the curve induced by the loads. The initial curve has coordinates x and y, and supports at ends x_0, y_0 and x_1, y_1. The arch deflects in the horizontal direction u and the vertical direction v. The slope dv/dx is designated i. The arch is subjected to a vertical load w and develops shears (vertical internal forces rather than true shears) F, bending moments M, and horizontal force H. The horizontal force does not vary along the length of the arch. Using the differential relationships between bending and deflection, he finds expressions for the shear, bending moment, slope, and vertical deflection at any point in the arch.

$$F = F_0 - \int_0^x w\,dx + H\left(\frac{dy}{dx} - \frac{dy_0}{dx_0}\right)$$

$$M = M_0 + F_0 x - \int_0^x \int_0^x w\,dx^2 - H\left(y_0 - y + x\frac{dy_0}{dx_0}\right)$$

$$i = i_0 - \int_0^x \frac{M}{EI}\sqrt{1 + \frac{dy^2}{dx^2}}\,dx$$

$$v = \int_0^x i\,dx$$

Some of the variables given are shown in Figure 6-2. F and H represent the vertical and horizontal internal force in the arch. The variable w represents the distributed load and may vary with x. The variable i represents the slope of the arch centerline, and v represents

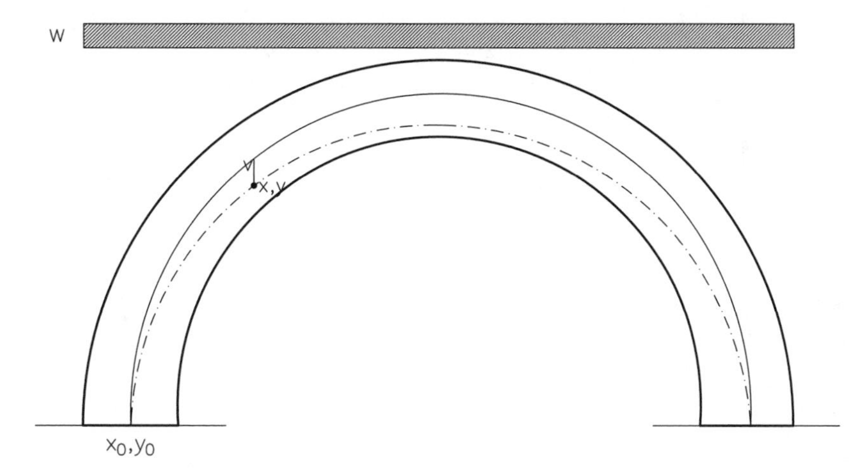

Figure 6-2. Rankine (1876b) nomenclature for masonry arch analysis.

the vertical deflection. In all cases a 0 subscript represents the value at the left support. Given four boundary conditions, these four equations constitute a formal solution of the unknown quantities, F_0, M_0, i_0, and H. From these four constants, the remainder of the internal forces can be found. The process is, however, tedious, and many of the integrals can only be explicitly evaluated in very limited cases.

A more practical analytical scheme is described by Malverd Howe (1897). Rather than trying to develop general analytical formulas for specific cases, Howe subdivides the arch ring and calculates the loads and relevant geometric properties at discrete points chosen in the subdivision of the arch ring into angular segments. The calculation procedure, although tedious, uses tabular computations, which are familiar to most engineers in the nineteenth century. Howe's advice on the construction of the tables is very specific and detailed. Using a discrete version of the same integral forms as Rankine, he finds the three unknown quantities H, M_1, and M_2 and the same support conditions for a fixed-fixed arch, for which the nomenclature is shown in Figure 6-3.

These conditions can be applied directly to a centerline divided into equal angular segments δs. Rather than imposing a specific load, Howe applies a unit load to each segment in turn and uses superposition to find the effect of actual loads. The computation of the horizontal thrust is accomplished in the first table (see Table B6-1-1), whereas the following table (Table B6-1-2) yields the support moments for unit loads at each of the segments. These quantities being computed, it is necessary to find the dead and live load influence on the remainder of the arch. This computational scheme did not present any particular difficulty to a nineteenth-century engineer. Howe (1914), in fact, recommends carrying the calculations to three decimal places by use of a machine or a Crelle (1897) table, asserting that this presents no particular difficulty compared with calculations with three significant figures. Howe then uses a semigraphical procedure to find the dead loads required to make the dead load thrust line follow exactly the center of the arch, so that only live load moments

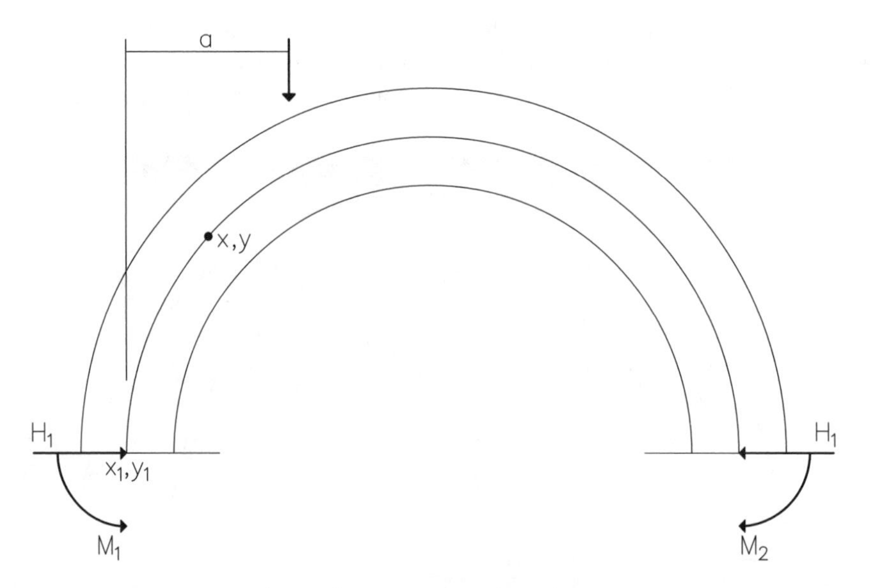

Figure 6-3. Howe (1897) nomenclature for masonry arch analysis.

need be computed. According to Howe, this is accomplished by modifying the density of the fill throughout the arch to remove the bending moments due to dead load from the arch. Because it is unlikely that an arch would be constructed this way (voids may be left over the haunches to reduce dead load but not in such a refined way as to calibrate the location of the dead load thrust line), this may be understood as an empirical dismissal of the importance of the moments due to dead loads in favor of calculating the response of the arch on the basis of the live loads only. Howe then finds the live load moments by superposition at the two supports and the crown, asserting that "If the ring is safe at these three points it will be safe at all other points." (p. 69) The entire process of calculating a segmental arch with a span of 60 ft and a rise of 7.5 ft is illustrated in Box 6-1.

Box 6-1

The treatment shown here is described in detail in Howe's *A Treatise on Arches* (1897), but is described in slightly simplified form and reduced to a tabular computation in the same author's *Symmetrical Masonry Arches* (1914, p. 76). In the preface to the more detailed work, *A Treatise on Arches,* Howe says that the tables were computed by method of differences, where possible, with checks by direct computation every tenth value. This is evidently the method of finite differences, as elaborated by Pearson (1849), which permits the analyst to fill in lines in the table based on the differences between previous entries, rather than completing more laborious computations. In Howe's development, the quantity m_x represents the bending moment at the abcissa x on a simply supported straight beam (see Figure B6-1-1). It is used to simplify complex expressions in the following development.

For an arch subjected to vertical load only, Howe uses the condition equations and expressions for the axial force and bending moment in the arch to write explicit expressions for the horizontal thrust and the bending moment at each support of an arch subjected to a single vertical load. The expressions used in the present development of

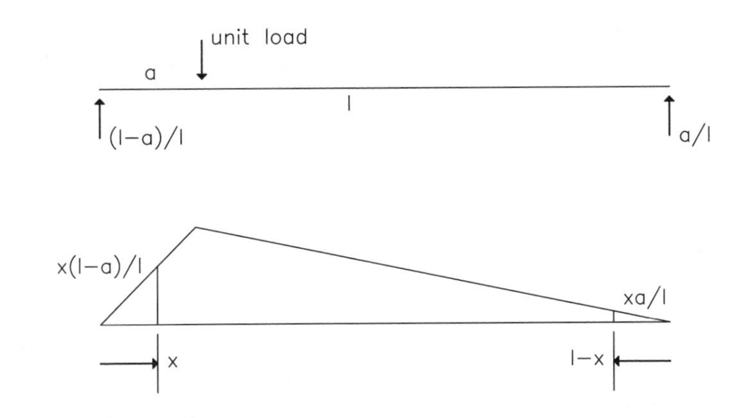

Figure B6-1-1. Calculation of m_x.
Source: Howe (1914).

the topic neglect the effect of deformation due to axial force and temperature change, although Howe later gives additional expressions that include these effects:

$$H = \frac{\Sigma m_x(y - y_a)}{\Sigma y(y - y_a)}$$

$$M_1 = Hy_a - \frac{\Sigma m_x}{n} - \frac{\Sigma m_x\left(x - \frac{l}{2}\right)}{\left(\frac{1}{2}nl - \frac{2\Sigma x^2}{l}\right)}$$

$$M_2 = Hy_a - \frac{\Sigma m_x}{n} + \frac{\Sigma m_x\left(x - \frac{l}{2}\right)}{\left(\frac{1}{2}nl - \frac{2\Sigma x^2}{l}\right)}$$

In these equations H is the horizontal internal force in the arch (neglecting the effect of axial compression and temperature change), m_x is the bending moment in a straight simply supported beam at the same x-coordinate, y_a is the average value of the y-coordinate of the centerline, M_1 and M_2 are the bending moments at the left and right supports, and n is the number of divisions in the full arch.

For example, this approach will be applied to a hypothetical arch with an 80-ft span, a 4-ft ring thickness, and a 10-ft rise. The first task, also completed in tabular form (Table B6-1-1), is to establish the x and y coordinates of the centerline of the arch ring. The table where these values are established is similar to Howe (1914, Table A, p. 75). The arch ring is divided into 20 equal segments; only one side of the axis of symmetry is represented. The radius of the intrados in this example is 85 ft, and the half-angle of embrace is 28°04′21″. The radius of the centerline is therefore 87 ft.

TABLE B6-1-1. *Arch Geometry*

Point	φ	sin φ	cos φ	R sin φ	R cos φ	x *40.9412-R* sin φ	y R cosφ-*75.00*
1	26°40′09″	.44883	.89361	39.005	77.744	1.936	2.744
2	23°51′43″	.40453	.91452	36.194	79.563	4.747	4.563
3	21°03′15″	.35925	.93324	30.385	81.191	10.556	6.191
4	18°14′49″	.31312	.94971	27.041	82.624	13.900	7.624
5	15°26′23″	.26623	.96391	23.162	83.860	17.779	8.860
6	12°37′57″	.21870	.97579	19.026	84.893	21.915	9.893
7	9°49′31″	.17065	.98533	14.846	85.723	26.095	10.723
8	7°01′05″	.12218	.99251	10.630	86.348	30.311	11.348
9	4°12′39″	.07343	.99730	6.388	86.765	34.553	11.765
10	1°24′13″	.02450	.99970	2.131	86.974	38.410	11.974
C	0	0	1.00	0	87.000	40.9412	12.000

To find H for a series of concentrated loads applied at points 1 through 10, it is necessary to find an expression for $\Sigma m_x(y - y_a)$ for this load case. The procedure used by Howe is to consider the bending moment diagram for a unit load placed at a (illustrated in Figure B6-1-1). The sum of the bending moment at x and $(l - x)$ is $[x(l - a) - xa]/l = x$, as the sum of the two reactions is the unit load. For moments at $a < x < (l - a)$, this sum is a constant a. As a result, for half the arch, the calculation of

$$\Sigma m_x(y - y_a) = \sum_{0}^{a} x(y - y_a) + a\sum_{a}^{l/2}(y - y_a)$$

by subtracting $\Sigma a(y - y_a)$, which is equal to zero because y_a is the average value of y over the domain, the simplest possible expression is obtained.

$$\Sigma m_x(y - y_a) = \sum_{0}^{a} x(y - y_a) - a\sum_{0}^{a}(y - y_a)$$

This expression is used in the tabular evaluation of H, displayed in Table B6-1-2. The end moments are calculated once H is known. As a preliminary to this calculation, it is necessary to determine Σm_x. By an argument similar to that used here, this is found to be

$$\Sigma m_x = \sum_{x=0}^{x=a} x + n'a$$

where n' is the number of divisions between the load and the crown. So this quantity is calculated in Table B6-1-3 for a unit load at points 1 through 10.

The equation given for M_1 and M_2 is now solved by introducing an exact expression for m_x and by simplifying as much as possible. The resulting equation is still somewhat complex and has the form shown as follows.

$$M_1 = Hy_a - \frac{\sum_{o}^{a} x + n'a}{n} - \frac{(R_1 - R_2)\sum_{0}^{a}\left(\frac{l}{2} - x\right)x + 2R_2\sum_{a}^{l/2}\left(\frac{l}{2} - x\right)^2}{\frac{4}{l}\sum_{0}^{l/2}\left(x - \frac{l}{2}\right)^2}$$

This equation is a slight modification of the equation previously given for M_1. In this equation R_1 and R_2 represent the left and right reaction on a straight simply supported bar with the same x coordinates as the arch under analysis. This equation is evaluated in Table B6-1-4 using a tabular calculation with 18 columns.

Howe balances the weight of the fill by an inverse procedure of drawing a concentric thrust line and finding the fill weights necessary to achieve this shape for the thrust line. He proposes modifying the density of the fill to achieve this effect. In this way, he dismisses any required dead load calculations. The dead load calculations for a constant density fill would be much less difficult than the calculations just completed, because an influence line for horizontal force and moment at the two supports has just been determined. In calculating the influence of the live load, Howe works with a

TABLE B6-1-2. *Determination of Horizontal Thrust*

	1	2	3	4	5	6	7	8	9	10	11
Point	Values of y	$y - y_a$	$y(y - y_a)$	x or a	$x(y - y_a)$	$\Sigma(y - y_a)x$	$-\Sigma(y - y_a)$	$a\Sigma(y - y_a)$	Columns (6) + (8)	H	Load at Point
0											0 or 0′
1	2.744	−5.8245	−15.984	1.936	−11.277	−11.277	5.825	11.277	0	0.000	1 or 1′
2	4.563	−4.0055	−18.279	4.747	−19.016	−30.293	9.830	46.663	16.37	0.093	2 or 2′
3	6.191	−2.3775	−14.722	10.556	−25.102	−55.395	12.208	128.867	73.472	0.417	3 or 3′
4	7.624	−0.9445	−7.205	13.900	−13.129	−68.524	13.152	182.813	114.289	0.667	4 or 4′
5	8.860	0.2915	−2.578	17.779	−5.174	−73.698	12.861	228.656	154.958	0.880	5 or 5′
6	9.893	1.3245	13.791	21.915	30.550	−43.148	11.536	252.811	209.663	1.191	6 or 6′
7	10.723	2.1545	23.097	26.095	56.209	+13.061	9.382	244.823	257.884	1.465	7 or 7′
8	11.348	2.7795	31.536	30.311	84.234	+97.295	6.602	200.113	297.408	1.689	8 or 8′
9	11.765	3.1965	37.601	34.553	110.431	+207.726	3.406	117.688	325.414	1.848	9 or 9′
10	11.974	3.4055	40.771	38.410	130.786	+338.512	0.00	0.00	338.512	1.923	10 or 10′
	85.685		88.028								
	2		2								
	171.370		176.056								
	$y_a = 8.5685$		$\Sigma(y - y_a)$								

Column 9 gives the value of the numerator of the formula for the calculation of H for each load. These values, divided by 176.056, the denominator, give the values of H shown in column 10.

TABLE B6-1-3. *Determination of* Σm *for Unit Loads*

1	2	3	4	5	6	
Division Point	x or a	$\sum_{x=0}^{x=a} x$	n'	$n'a$	Col 5. + Col. 3 Σmx	Load Unity at Point
1	1.936	1.936	9	17.424	19.360	1 or 1′
2	4.747	6.683	8	37.976	44.659	2 or 2′
3	10.556	17.239	7	73.892	91.131	3 or 3′
4	13.900	40.139	6	83.400	123.539	4 or 4′
5	17.779	57.918	5	88.895	146.813	5 or 5′
6	21.915	69.833	4	87.660	157.493	6 or 6′
7	26.095	95.928	3	78.285	174.213	7 or 7′
8	30.311	126.239	2	60.622	186.861	8 or 8′
9	34.553	160.792	1	34.553	195.342	9 or 9′
10	38.410	199.212	0	0	199.212	10 or 10′

movable load in lbs/ft, say 400 lbs/ft, and finds the corresponding maximum forces for live loads.

Howe chooses the support points, the crown, and point 6 ft to investigate maximum moments due to variable load placement. For instance, having determined (somewhat arbitrarily) that for the maximum negative moment at the crown, loads at 1 through 7 and 1′ to 7′ produce negative moments, he calculates the sum of the effect of a unit live load at each of these points, based on a tabular computation using the formula:

$$M_x = M_1 + V_1 x - Hy - P(x - a)$$

In this formula x represents the point at which moments are being calculated, namely at the crown, $x = 40.94$ ft, and a represents the x coordinate of the point where the individual loads are applied. The vertical reaction at the left support V_1, is calculated from the values in Table B6-1-4, by the equation

$$V_1 = R_1 + \frac{(M_2 - M_1)}{l}$$

Following the example here, the effect at the crown is twice the effect of unit loads at points 1 through 7, which can be calculated according to Table B6-1-5. In this table, the variables M_1, M_2, R_1 (used in finding V_1) and H are determined from Table B6-1-4. The values used in the table relate to the unit loads at point 1 through 7 until the point where a load of 400 lb/ft times the tributary area of each point is substituted. As a result, the negative moment at the crown caused by loading at points 1 through 7 and 1′ through 7′ is equal to 2 × 8,986 ft-lb = –17,970 ft-lb. A check of loads 1 through 6 or 1 through 8 would assist in verifying that the live load placement selected produces the maximum moments.

TABLE B6-1-4. *Determination of End Moments for Unit Loads*

	1	2	3	4	5	6	7	8	9	10	11	12	13	14	15	16	17	18
Point	R_1	R_2	x	$l/2\text{-}x$	$x(l/2\text{-}x)$	$(l/2\text{-}x)^2$	$\sum_0^a x\left(\frac{l}{2}-x\right)$	$R_1\text{-}R_2$	Column 7 × Column 8	$\sum_a^{l/2}\left(\frac{l}{2}-x\right)^2$	$2R_2$	Column 10 × Column 11	Column 9 × Column 12	Column 13 divided by denominator	$\Sigma m_x \div n$	Hy_a	M_1	M_2
1	0.976	0.024	1.936	38.064	73.692	1,448.868	73.692	0.952	70.125	3,931.197	0.048	190.270	260.395	0.961	0.968	0.000	−1.929	−0.007
2	0.941	0.059	4.747	35.253	167.346	1,242.774	241.038	0.881	212.433	2,688.423	0.119	319.049	531.481	1.961	2.233	0.797	−3.397	0.525
3	0.868	0.132	10.556	29.444	310.811	866.949	551.849	0.736	406.216	1,821.474	0.264	480.687	886.903	3.272	4.557	3.573	−4.256	2.289
4	0.826	0.174	13.900	26.100	362.790	681.210	914.639	0.653	596.802	1,140.264	0.348	396.242	993.043	3.664	6.177	5.715	−4.126	3.202
5	0.778	0.222	17.779	22.221	395.067	493.773	1,309.706	0.556	727.574	646.491	0.444	287.349	1,014.923	3.745	7.341	7.540	−3.545	3.944
6	0.726	0.274	21.915	18.085	396.333	327.067	1,706.039	0.452	771.343	319.424	0.548	175.004	946.347	3.492	7.875	10.204	−1.162	5.821
7	0.674	0.326	26.095	13.905	362.851	193.349	2,068.890	0.348	719.198	126.075	0.652	82.248	801.446	2.957	8.711	12.552	0.885	6.798
8	0.621	0.379	30.311	9.689	293.683	93.877	2,362.573	0.242	572.274	32.198	0.758	24.399	596.673	2.201	9.343	14.471	2.927	7.330
9	0.568	0.432	34.553	5.447	188.210	29.670	2,550.783	0.136	347.353	2.528	0.864	2.184	349.537	1.290	9.767	15.834	4.777	7.356
10	0.520	0.480	38.410	1.590	61.072	2.528	2,611.855	0.040	103.821	0.000	0.960	0.000	103.821	0.383	9.961	16.476	6.133	6.899
						sum = 6,225												

TABLE B6-1-5. *Determination of Crown Moment for Distributed Live Load*

Load at point	x or a	y	M_1	V_1x (x = 40.94)	Hy	$(x - a)$	Live Load	M
1	1.936	2.744	−1.929	39.992	0	38.064	1,936	0
2	4.747	4.563	−3.397	39.587	1.116	35.253	1,724	−308
3	10.556	6.191	−4.256	37.994	5.004	29.444	1,831	−1,299
4	13.900	7.624	−4.126	36.713	8.004	26.100	1,445	−2,190
5	17.779	8.860	−3.545	34.855	10.56	22.221	1,603	−2,358
6	21.915	9.893	−1.162	32.534	14.292	18.085	1,663	−1,671
7	26.095	10.723	0.885	29.909	17.58	13.905	1,679	−1,160
							ΣM	−8,986

Semigraphical Methods

Semigraphical methods depend on the construction of a thrust line and are focused on determining the proper placement of the thrust line and other characteristics. As described later in the book, the simplest method for constructing the thrust line is graphical. Additional clarification on the graphical construction of the thrust line is available in Chapter 13. The thrust line also can be constructed analytically by taking the bending moment at any point and dividing by the horizontal thrust to obtain a y coordinate.

The methods described in Baker (1907, Chapter 18) are procedures for determining the correct location of the line of pressure, with the understanding that it is possible to draw a line of pressure under any reasonable set of assumptions. Ira Osborn Baker (1907) describes and George Fillmore Swain (1896) presents some important terms relative to the description of failures in the arch. The line of resistance is defined as the locus of centers of pressure of the resistance to internal weights and external forces in the arch. Criteria for the stability of the arch generally are given in terms of the line of pressure. Swain's criteria for stability are as follows.

- The true line of resistance must be within the arch ring, or within the middle third if there is to be no tension.
- The true pressure on any joint must not make with the normal to that joint a greater angle than the angle of repose.
- The maximum intensity of pressure at any joint must not exceed the allowable stress.

Baker also introduces what is now known as Heyman's (1995) geometrical factor of safety in which a rough factor of safety against collapse of the arch is determined as the quotient between the half-thickness of the arch and the deviation of the line of resistance from the center.

The theories that Baker describes are these:

Least Crown Thrust. By locating the line of pressure as high in the arch as possible, the horizontal thrust at the crown, and consequently the thrust throughout the arch, can be minimized. The further requirement of this method is to ensure that the thrust line passes

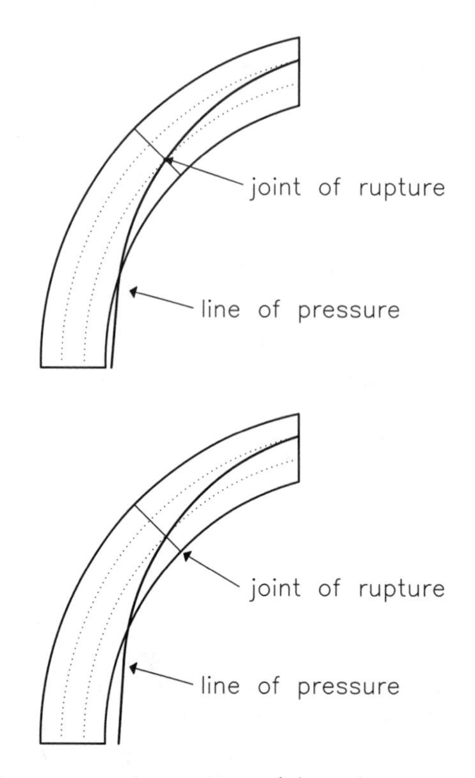

Figure 6-4. Various hypotheses concerning position of thrust line.

through the intrados at the joint of rupture, as the selection of the joint of rupture gives the extent of the arch that needs to be considered (see Chapter 2). The level of the horizontal thrust to be considered is also limited if the designer adopts the rule of the middle third, in which the center of pressure at an arch joint has to be contained within the middle third of the thickness of the arch ring. This is illustrated in Figure 6-4, where the thrust line associated with the least pressure is found to pass through the extrados (or the outer limit of the middle third) at the crown and at the abutment and through the intrados (or the inner limit of the middle third) at the joint of rupture.

Least Pressure. In the least pressure method, a thrust line is drawn for which the pressure (maximum compressive stress due to axial force + moment) in any joint is the minimum consistent with equilibrium. This generally means keeping the thrust line as close to the center of the arch as possible. A. Jay Du Bois (1887) espouses this theory, although Baker (1907) dismisses this theory as having no rational basis. Swain (1896) and Baker (1907) differ over the importance of the hypothesis of least crown thrust. While Baker says it is the most frequently employed for the determination of the true line of resistance in the arch, Swain ignores it in favor of a graphical and incremental construction of the thrust line lying closest to the centerline of the arch.

Scheffler's Theory. Scheffler's theory is an elaboration of the least crown thrust theory in which the absolute minimum thrust is determined by summation of moments through the joints from the crown downward, with the center of rotation considered to be the upper middle third line of the joint.

Rational Theory. In the rational theory, which seems to be a composite of the aforementioned theories adopted by Baker, both vertical and horizontal forces are accounted for at each joint. The joint of rupture is determined. This is done by a tabular computation, similar to that shown in Box 6-1. The horizontal thrust is calculated at each joint starting at the crown, assuming the horizontal crown thrust is placed at the upper end of the middle third. Following this, the horizontal thrust is calculated on the basis of the thrust line passing through the lower end of the middle third at each successive joint. The joint for which the calculated value of the horizontal thrust is a maximum is the joint of rupture. On the basis of the joint of rupture, a thrust line is drawn by accepted graphic methods on the basis of least crown thrust determined in this manner.

Winkler's Theorem. According to Winkler's theorem, the correct position of the thrust line is that in which the least squares sum of the deviations from the arch centerline are at a minimum. Although simple enough in statement, the application of this theorem requires drawing a thrust line, measuring the error at each joint, calculating the sum of the squares, and repeating the process until a satisfactory minimum is found.

Later authors, particularly Jacques Heyman (1995), have identified the construction of lines of pressure with the limit theorems of plastic analysis and have determined that the construction of a single statically admissible line of pressure, entirely within the arch, is sufficient to ensure stability. Baker (1907, p. 464) disputes the widely held view that a corollary of Winkler's theorem is that a statically admissible thrust line lying within the middle third implies that the true elastic thrust line is also within the middle third, although he does not express an opinion on Scheffler and Fournié's (1864) similar claim for the middle half of the arch.

In general, analytical methods for arch analysis presented formidable but not insurmountable obstacles to successful analysis by the methods of nineteenth-century engineering. Although the mechanics of the response of an elastic arch were well understood, their application in the case of the variable geometry and statical indeterminacy of an arch made it extremely tedious to complete a full analytical solution. Even discretizing the problem and solving by approximate methods (as Howe has done) results in a complex analysis performed on various large tables. Recourse to semigraphical methods was often taken, that is, the determination of possible lines of pressure by graphical methods with the determination of the correct line of pressure by analytical methods. Empirical methods, almost all developed on the basis of analyzing populations of analytically design arches, described in Chapter 2, were very often employed. For more complex structures, it was usual to make use of graphical methods, such as those described in Chapter 13.

References Cited

Baker, I. O. (1907). *A treatise on masonry construction*, 9th Ed. John Wiley and Sons, New York.

Crelle, A. L. (1897). *Dr. A. L. Crelle's calculating tables giving the products of every two numbers from one to one thousand and their application to the multiplication and division of all numbers above one thousand.* David Nutt, London.

Du Bois, A. J. (1887). *The strains in framed structures*, 4th Ed. John Wiley and Sons, New York.

Gauthey, E. M. (1771). *Mémoire sur l'application des principes de la mécanique à la construction des voûtes et des domes.* Louis-Nicolas Frantin, Dijon.

Haupt, H. E. (1856). *General theory of bridge-construction.* D. Appleton, New York.
Heyman, J. (1995). *The stone skeleton.* Cambridge University Press, Cambridge, MA.
Howe, M. A. (1897). *A treatise on arches*, 1st Ed. John Wiley and Sons, New York.
Howe, M. A. (1906). *A treatise on arches*, 2nd Ed. John Wiley and Sons, New York.
Howe, M. A. (1914). *Symmetrical masonry arches.* John Wiley and Sons, New York.
Kidder, F. (1886). *The architect's and builder's pocket-book*, 3rd Ed. John Wiley and Sons, New York.
Pearson, J. (1849). *The elements of the calculus of finite differences.* E. Johnson, Cambridge.
Rankine, W. J. M. (1876a). *A manual of civil engineering*, 11th Ed. C. Griffin, London.
Rankine, W. J. M. (1876b). *Manual of applied mechanics*, 8th Ed. C. Griffin, London.
Scheffler, H., and Fournié, V. (1864). *Traité de la stabilité des constructions.* Dunod, Paris.
Swain, G. F. (1896). *Notes on the theory of structures*, 2nd Ed. Mimeographed lecture notes, Cambridge, MA.

7

Analysis of Braced Girders and Trusses

In the usage of the nineteenth century, a truss may be considered a girder when it is used as the main support of a bridge span. As such, we will consider the analysis of trusses in this chapter, but we will include continuous trusses in the study of the analysis of continuous girders to be found in Chapter 8. This terminology follows the development of trusses, which originally functioned more like a girder with a braced web (Figure 7-1), than the modern sense of a truss. In the evolution of the forms of these girders, the flanges became more attenuated as the bars began to be considered as only axially loaded. The final result was the forms of truss that are more familiar to us and the forms of analysis in which trusses and solid-web girders are differing structural forms. The design of wood trusses by the application of rules, as propounded by Thomas Tredgold (1888), Robert Griffith Hatfield (1871), and others is covered in Chapter 3.

Three analytical methods prevail for trusses in the later nineteenth century. The first is the development of formulas that cover the forces in the chords, ties, and struts, in terms of the panel load, panel length, number of panels, and so on. John Davenport Crehore (1886) presents a variety of means of analyzing trusses, however, his work exemplifies the approach of developing formulas specific to a truss type. Crehore divides trusses into 12 classes and presents formulas for the forces in the bars of each of the classes, when the truss is subject to different forms of loading. George Fillmore Swain (1896) also develops similar formulas for certain classes of truss bridge, especially those with curved chords. The second is the application of the method of moments or of formulas derived from the method of moments. This method is used especially by A. Jay Du Bois (1888), but also by Swain

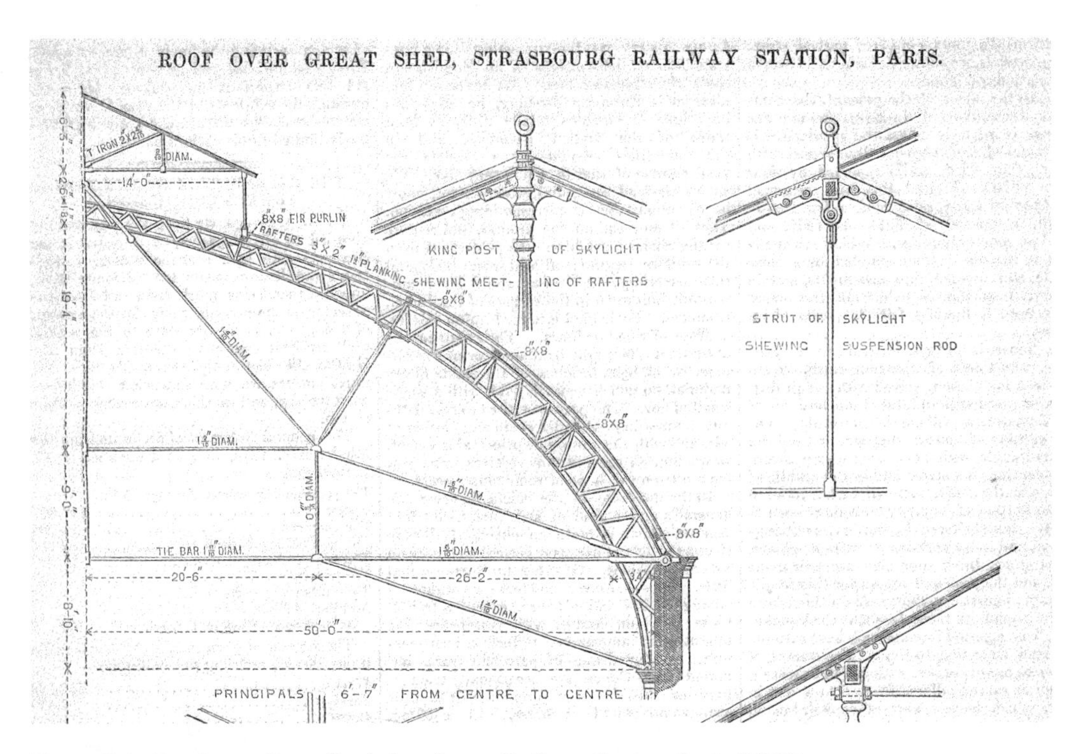

Figure 7-1. Roof over Great Shed, Strasbourg Railway Station, Paris (1850).
Source: n.a. 1850.

and others. The third is the method called indexing by Swain; in this method, the forces are traced through the truss by rapid analysis of the overall shear forces. Du Bois presents a variant of this method, and it is further available in Charles Lee Crandall's work (1888) and the work of other authors.

Two other methods of analysis are deferred to a later chapter. Trusses can be examined graphically by constructing scaled diagrams representing the forces in each of the bars. Also available are semigraphical methods in which the forces in the truss loaded at a single panel point are determined graphically, and forces from loads at different panel points are summed analytically. Both graphical and semigraphical methods are described in Chapter 12.

The earliest bridge trusses in the United States certainly met the description of braced girders. Neither the Bollman truss nor the Fink truss (Figure 7-2) actually has a bottom chord. The two forms represent adaptation of the queen-post truss, already used for static loads on roofs, to the movable loads present on bridges. Each panel point is made into a loaded point and a tension cable is used for the support of each panel point.

One of the earliest codifications of methods for the analysis of trusses was by Robert Henry Bow (1874, p. 8). He discusses the distinction between *trussing*, as used in the web of a gable truss with continuous chords, that is, a structure that does not require additional stability, and *bracing*, the insertion of additional members to ensure the stability of a girder. Bow's four classes are parallel chord, nonparallel chord, arches, and braced double arches. Considering the inverted truss, with a constant load w imposed at each upper panel point,

(A) (B)

Figure 7-2. Bollman truss, Savage County, MD (HAER MD, 14-SAV,1–13). Fink truss, Clinton County, NJ (HAER NJ, 10-CLIN.V,1–1J).
Sources: (A) Photograph by William E. Barrett. (B) Photograph by Jack E. Boucher.

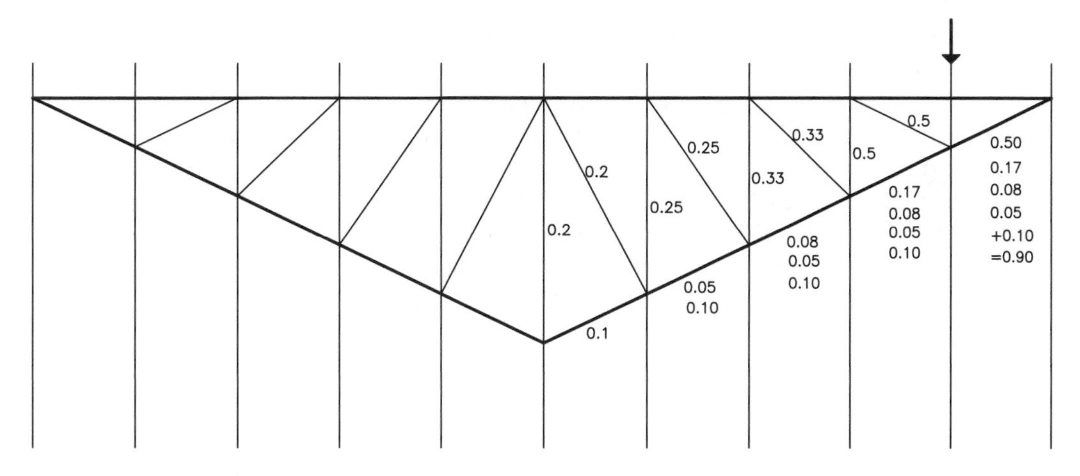

Figure 7-3. First stage of Bow's analysis of an inverted gable truss.

the distribution of the weight applied at this point can be traced through the remainder of the truss. An example of Bow's analysis of an inverted, pitched bottom chord truss is shown in Figure 7-3. The truss in Figure 7-3 is a variant on the types of truss analyzed by Bow using this method.

In Bow's analysis, the vertical load applied at the second panel is divided into two parts: the bottom chord and the diagonal in the second panel, 0.5 being written over the web diagonal and the bottom chord in the first panel. The 0.5 force in the second panel is transmitted through the second web vertical, where the remaining force is divided two-thirds to the web member and one-third to the bottom chord, based on the slopes of these members: 0.33 is written over the web member as the vertical component of force, 0.33 is written over the web diagonal, and 0.17 is written over the bottom chord in the second panel. Because this force of magnitude 0.17 is equally transmitted into the bottom chord at the first panel, it is written over that member as well. The vertical force of 0.33 is similarly subdivided at the third panel point: three-fourths to the web and one-fourth to the bottom chord, and the result for the bottom chord is written under each part of the bottom chord to the support. This process is continued to mid-girder, and the resulting force in the bottom chord is the sum of all of the vertical components transformed to a resultant force by the slope of the bottom chord. The result of this analysis is a complete analysis of the truss for a single load, resulting in the correct reaction placed at the right support. The sum of the numbers over the top chord is the vertical component of the chord force (similarly for the diagonal web members).

This analysis was adapted by Bow (1874) to include a truss with every panel loaded. The diagonal forces also can be converted to compressive forces by multiplying by the secant of the angle each diagonal makes with the horizontal, resulting in a diagram (Figure 7-4) similar to Bow's Figure 56. On the left is shown the distribution of the loads to the web and the summation of the vertical components of the bottom chord segments. The actual forces are shown on the right half of the truss. The bottom chord forces shown result from a choice of slope that makes these forces exactly twice their vertical component, that is, a slope of 30 degrees.

In discussing parallel chord trusses, which he calls Class the First, Bow takes a similar approach to the determination of the vertical component of the forces in a Warren girder. The proportion of the load at each point sustained by each of the supports is simply determined by the lever rule (left and right reactions are proportional to distance to right and left supports). If all the panel points are loaded, the total force in each brace and in the girder can be determined by combination of the force effects at the various points or by writing out formulas for the vertical force component at each point (later known as index stresses). Thus, for instance, for a truss of N panels, the maximum panel shear force at panel n for a unit load applied to every panel is $(N\text{-}n)^2/2N \times w$, where w is the panel load.

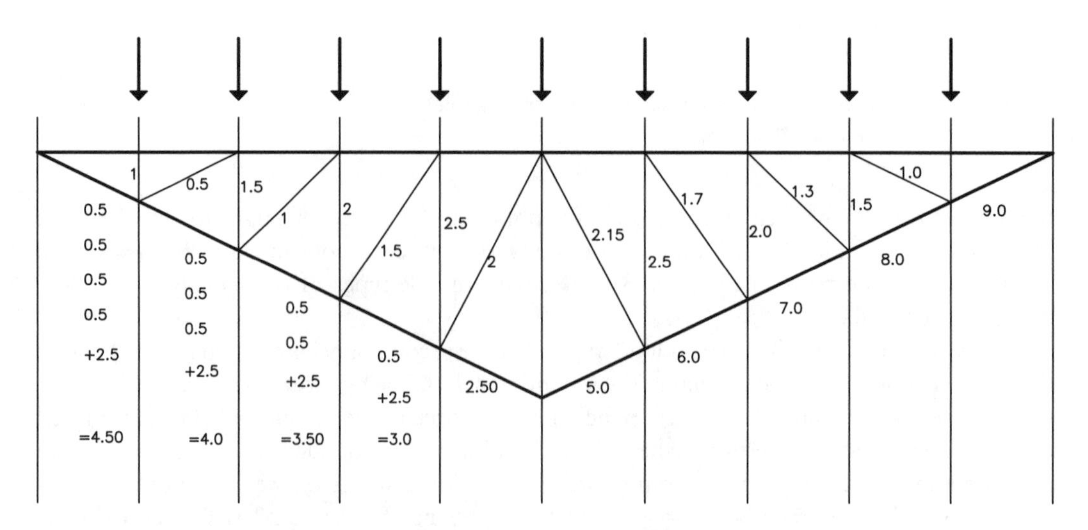

Figure 7-4. Summation of loads and forces on an inverted gable truss.

Many of Bow's methods were extended and completed by Crehore (1886). Crehore chose a much wider range of classes of truss and developed usable formulas for the force in the bars of trusses of various classes, rather than any attempting to write a general analysis method for all trusses. He develops the basic method of computing horizontal force components from moments (p. 64) and the further procedures of using differences in horizontal force components in a panel to find the horizontal component of the force in a web bar. Formula for Class II (only bottom chord horizontal, p. 106) is a typical example of this treatment. These procedures will be discussed later in this chapter as part of the discussion of curved chord trusses. Crehore is also specific about the resolution and superposition of multiple systems. He presents explicit formulas for moments at a point due to various patterns of loading tables (p. 33 has a good initial example) for combinatorial values of multiple loadings on trusses (for instance, pp. 425–428). An example of one of these tables is reproduced in Figure 7-5.

Olaus Henrici (1867) describes the construction and analysis of "skeleton structures" that are, in fact, trusses. In his comments, he shows a clear preference for statically determinate forms (p. 13). He states, "This example also shows clearly that in any structure, if there are more bars than necessary to determine the figure, they are not only superfluous, but injurious and dangerous, because, not knowing the exact amount of error in each, it is

150. To compute the Moments and Strains in Chords due to the Total Panel Weight, $W + L$, applied at Each Apex, Top and Bottom.— We have, from equation (65), moments at the vertical sections through these apices,

$$M_r = \frac{W + L}{2n} l(n - r)r = \frac{W + L}{2n} l\varepsilon_4. \qquad (480)$$

VALUES OF ε_4 IN (480).

r	n = 4	6	8	10	12	14	16	18	20	22	24
1	3	5	7	9	11	13	15	17	19	21	23
2	4	8	12	16	20	24	28	32	36	40	44
3		9	15	21	27	33	39	45	51	57	63
4			16	24	32	40	48	56	64	72	80
5				25	35	45	55	65	75	85	95
6					36	48	60	72	84	96	108
7						49	63	77	91	105	119
8							64	80	96	112	128
9								81	99	117	135
10									100	120	140
11										121	143
12											144

Figure 7-5. Calculation table for trussed girder moments.
Source: Crehore (1886).

not possible to ascertain the exact strain by calculation." He proceeds to demonstrate the calculation of forces in bars of statically indeterminate trusses by solving for equilibrium at successive joints (pp. 25–33) and advocates (p. 33) proportioning the structure so that the stresses in each bar are at a maximum. For statically indeterminate structures with superfluous bars, he takes a simple example derived from a suspended chain truss (p. 40) and writes equilibrium equations into which he substitutes displacement quantities for the bar forces. This permits him to solve for the bar forces. Having found bar forces based on equal bar areas, he chooses new bar areas based on the principle of having each bar at a maximum stress level. Solving the same equations again for the bar forces gives a new set of bar areas, resulting in an iterative procedure that increases the economy of the truss over a constant cross-sectional area solution.

Other authors, including Hermann Haupt (1856, p. 79–81), Francis Webb Sheilds (1867), and Bindon Blood Stoney (1873, pp. 87–164) write on the analysis of trusses. Haupt presents a graphical analysis of trusses, similar to that shown in William Merrill (1870). These methods are discussed in Chapter 12 under the graphical analysis of trusses. Sheilds develops formulas very much like those of Bow or Crehore, and Stoney reviews the significant details of the construction of a truss bridge, the Boyne Lattice Bridge in an appendix. Haupt presents analysis methods for wooden Howe trusses. The methods are separated into analysis/design of the chords, bracing, and counter bracing. Some of the analysis methods for the chords are taken up more generally by Merrill, described in Chapter 12. Haupt's analysis of the braces is highly simplified in that he simply applies panel loads to each brace and determines the brace force resulting from the panel loads by trigonometry. He does describe an interesting statically indeterminate type of truss, a wooden Howe truss with "arch braces," that is, braces that extend directly from the abutment to the top chord. He says that this type of bracing has been added to several sagging bridges. In this case, he advises that the entire bridge weight should be carried by the arch braces, with the panel braces carrying only the load of the individual panel that they are supporting.

Indexing Methods for Parallel Chord Trusses

The method described in this section is based on the writings of Swain (1857–1931) but is evident in the works of other authors such as Bow (1874), Crandall (1888), and others. Swain was an 1877 graduate in civil engineering from the Massachusetts Institute of Technology (MIT), where he served as department head from 1887 to 1909. He finally published his lecture notes on structural engineering in three volumes in 1927. The lecture notes themselves were used in his courses at MIT (first edition dated 1892, second edition dated 1896). In his discussions on the design of truss bridges, Swain consolidated several ideas, visible in the work of other authors, notably Bow, Merrill, and Du Bois, on the rapid analysis of parallel chord truss bridges. The result was a method titled "indexing," in which the analyst keeps track of panel shears caused by the loading and converts these panel shears to bar forces when necessary. The indices represent the vertical panel shear for the loading under consideration.

The principle of indexing is demonstrated by a simple six-panel Pratt truss, illustrated in Figure 7-6. Panel length is p and truss height is h, while l denotes the length of a diagonal. The panel load P for each is calculated as the product of panel length, half the floor width,

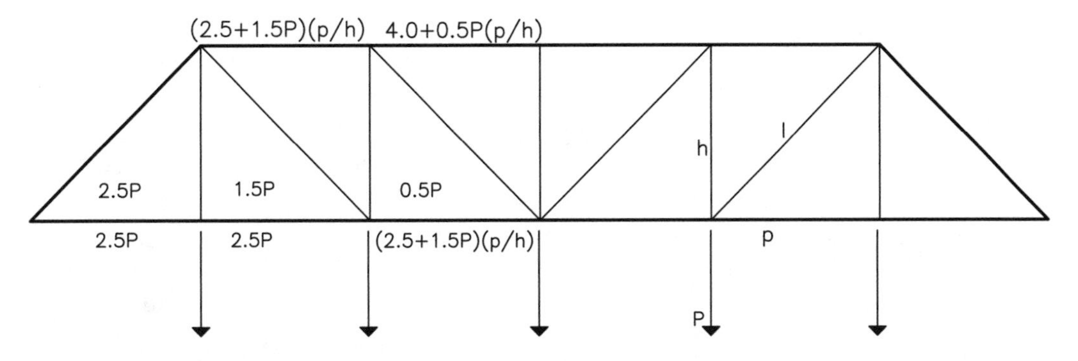

Figure 7-6. Illustration of the principle of indexing: 6-panel Pratt truss.

and the uniformly distributed floor load. Looking at an elementary example of a six-panel Pratt truss bridge, the support reactions under full load are $2.5P$. The panel shear in the first, second, and third panels is, respectively, $2.5P$, $1.5P$, and $0.5P$. As a result of this loading, the force in the braces in each panel is $2.5P(l/h)$, $1.5P(l/h)$, and $0.5P(l/h)$. Because the horizontal component of the brace forces [$2.5P(p/h)$, $1.5P(p/h)$, and $0.5P(p/h)$] combine in the top and bottom chords, the top and bottom chord forces can be represented as sums of panel shears, starting from the support, as illustrated in the figure.

For a conventional truss form, the strain sheet could be developed in a few minutes by an experienced designer, working from a scaled drawing of the truss and a calculation of the panel loads, simply by keeping track of the vertical component of the panel shears. The loading applied to road bridges was particularly simple, expressed as a "rolling load" in lbs/ft^2 of deck, which can be easily transformed to a panel load. As an example, consider the eight-panel single intersection pin-connected Pratt Truss, depicted as Design "P" in the Phoenix Bridge Company *Album of Designs* (1888), reproduced in Figure 7-7. The bar forces and indices developed in this discussion are illustrated in Figure 7-8. A similar bridge, designed by the Phoenix Bridge Company (with seven panels instead of eight) is shown in Figure 7-9.

The dead load can be estimated from several sources, including the tables of bridge dead loads published by John Alexander Low Waddell (1894), or using the formulas for bridge dead load published by Frank Oliver Dufour (1909). According to Waddell, Table I, a 180-ft span Class A Pratt Truss bridge with 20-ft roadway has a weight of 957 lb/ft, which is here rounded to 500 lbs/ft/truss.

Span:	179 ft – 8 in.	
Width:	21 ft – 8 in. + 25 ft – 0 in. sidewalks	
Panel length:	22 ft – 5½ in.	
Height:	26 ft – 0 in.	
Loading:	Weight:	500 lbs/ft/truss
	Rolling load:	60 PSF (from *Album of Designs* (p. 19) for bridge > 120 ft)
	Sidewalk load:	60 PSF (from *Album of Designs*)
Panel load:	Dead:	5.61 tons/truss
	Live:	10.7 tons/truss
	Total:	16.3 tons/truss

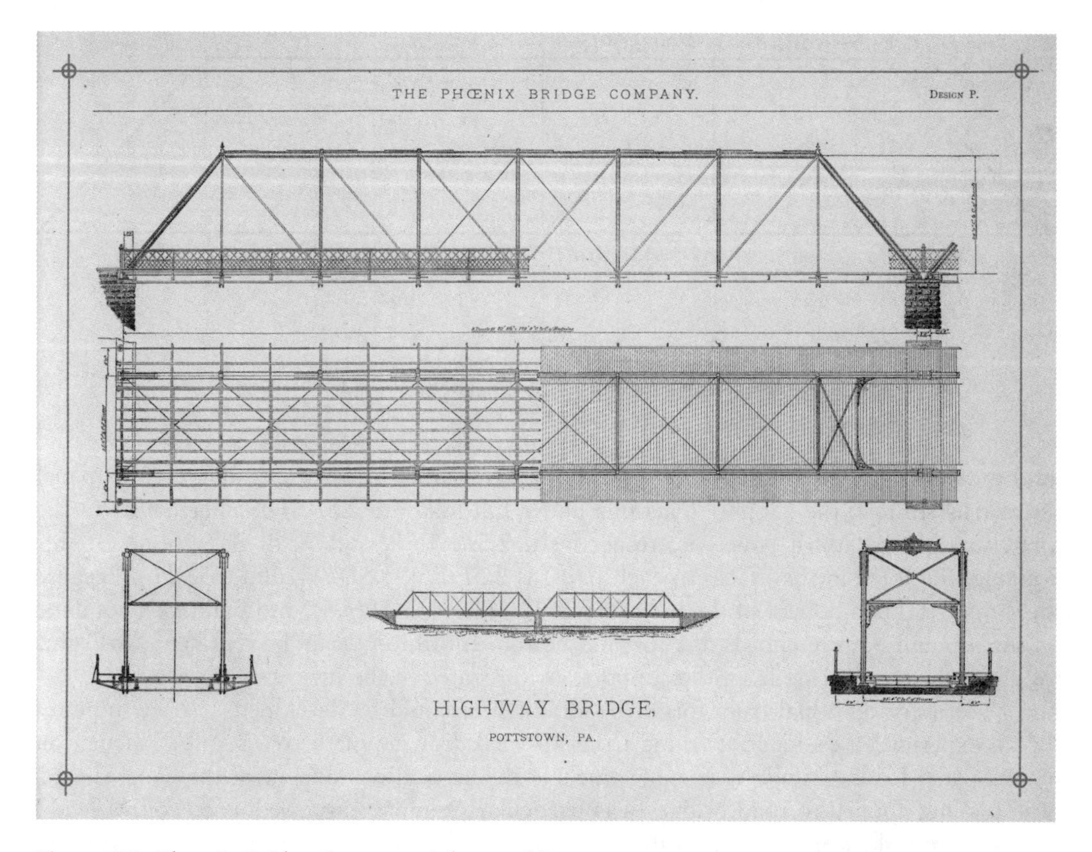

Figure 7-7. Phoenix Bridge Company eight-panel Pratt truss.
Source: J. B. Lipincott (1988).

Because the bar forces depend only on the number of panels, with the *p*/*h* and the *l*/*h* ratios as variables, it is possible to develop a generic table of the bar forces. Waddell, among others, published tables of forces for single-intersection and double-intersection Pratt trusses. To find the maximum force in a bar of a truss of a given type, it is necessary to correct a tabulated value for the panel length/height (*p*/*h*) ratio of the truss, then multiply by the panel length and load. An example of the application of Waddell's (1894) table (shown in Figure 7-10) for the truss in Figure 7-8 is shown following. In this truss tan θ = panel length/height (*p*/*h*) = 0.864, and sec θ = brace length/panel height (*l*/*h*) = 1.32.

In this table (Figure 7-10), the member in the bottom chord adjacent to mid-span is numbered Bottom Chord Member 4 and has an index force of 7-1/2 × (dead load + live load) = 7.5 (16.3 tons) = 122 tons, which, when multiplied by the panel length/height ratio of 0.864 = 106 tons. For web members, it is necessary to use a different factor for dead and live loads. For instance, the web diagonal in the third panel is labeled "diagonal 2" and has index stresses of 15/8 live load and 1-1/2 dead load. The force in the brace is then equal to 10.7 (15/8) + 5.61 (1.5) = 28.5 tons. The maximum force in the brace is the index stress

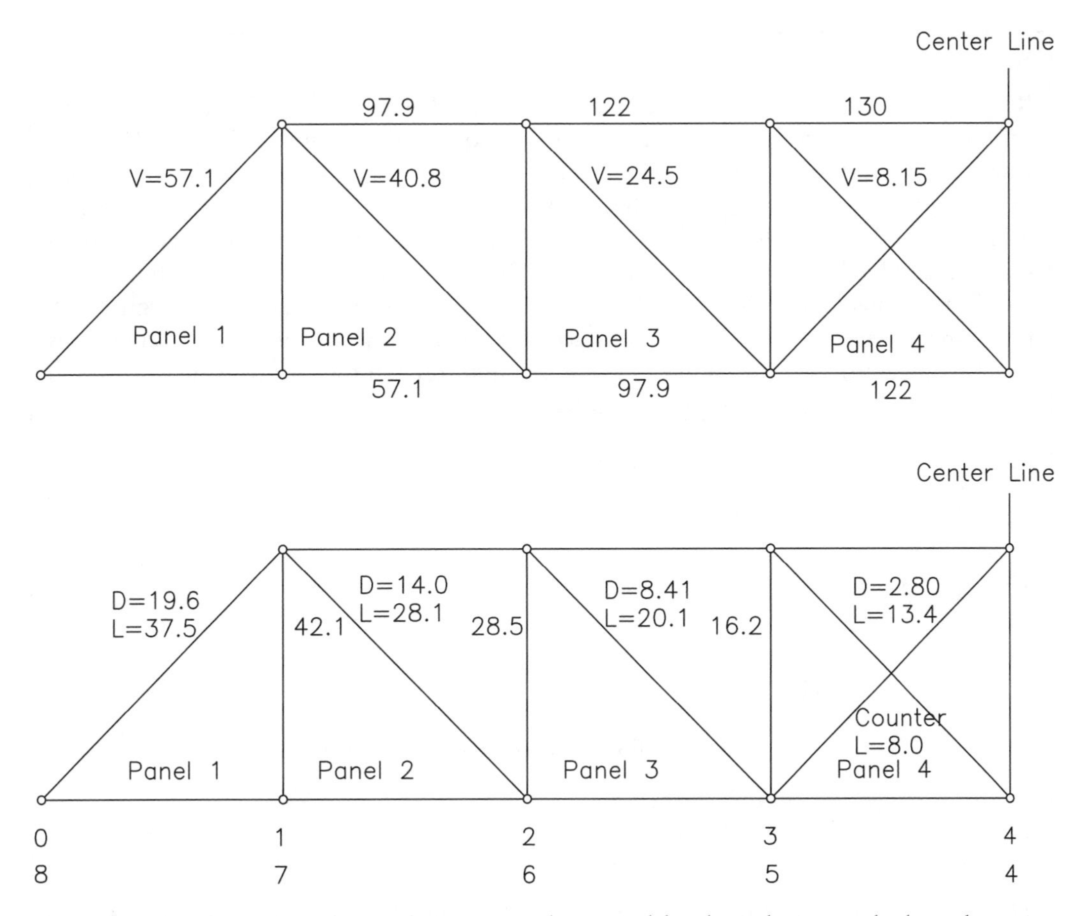

Figure 7-8. Bar forces in eight-panel Pratt truss determined by the indexing method; to determine actual chord forces, multiply indices by p/h (panel length/height) ratio 0.863.

multiplied by the brace length/panel length ratio, thus 37.6 tons. In Box 7-1 is described the full application of the indexing method for the determination of bar forces in a single intersection and a double intersection Pratt truss. The panel count numbers written below the panel points in the lower part of Figure 7-8 are used in the boxed discussion for the calculation of forces in the web members

These numbers may also be used in a different way for the determination of forces in the bars of a truss. Adapted to a parallel chord truss, it is possible to use the subscripted panel counts to determine the force in the chords. For a load per panel P and a panel length p, the moment about any of the panel points is equal to half of the product of the upper and the lower index multiplied by Pp. Then the mid-span moment in the Phoenix truss in Figure 7-8 is

$$M = 1/2(4 \times 4)16.3 \text{ tons} \times 22.46 \text{ ft} = 2930 \text{ ton-ft}$$

Then the fully loaded force in the top chord is this moment divided by the truss height

$$F = 2930 \text{ ton-ft} \div 26 \text{ ft} = 113 \text{ tons}$$

Box 7-1

Refer to Figure 7-8 for the results of this calculation. The *p/h* ratio of horizontal to vertical force components is established by the slope of the struts and ties: 22′-5 1/2″ horizontal to 26′ vertical, or 0.863 horizontal to 1 vertical. The *l/h* ratio of diagonal force in a panel to vertical force is 1.32. Maximum top and bottom chord forces are determined using full dead and live load, which are expressed as their vertical components, based on the total dead + live panel load of 16.3 tons. Because this truss has an even number of panels, the panel load is divided in half at the axis of symmetry at mid-span to give 8.15 tons for the center panel. The vertical component of strut and tie forces accumulates panel by panel toward the support. The horizontal force in the chords is first expressed as a sum of vertical panel component forces.

Panel 1 has a horizontal component of

$$57.1 \times 0.863 = 49.3$$

The top chord force in Panel 2 is

$$(57.1 + 40.8) \times 0.863 = 84.5 \text{ tons}$$

In Panel 3, the top chord force is

$$(57.1 + 40.8 + 24.5) \times 0.863 = 106 \text{ tons}$$

Finally, the top chord force in Panel 4 is

$$(57.1 + 40.8 + 24.5 + 8.15) \times 0.863 = 113 \text{ tons}$$

The bottom chord force in Panel 1 and 2 is 49.3 tons, in Panel 3 is 84.5 tons, and in Panel 4 is 106 tons. The forces are left in index (vertical component) form on the diagram in Figure 7-8 to allow their direct addition on the diagram. The eventual solution of the forces involves multiplying all of the indices by a fixed number *p/h* and *l/h* ratio. This final conversion can be completed with one setting on a slide rule.

The maximum force in the ties and struts is established by placing the live load in all the panels up to the panel in which the tie and strut in question are located, whereas the dead load must remain on every panel. The procedure begins by numbering each of the panel points 1 through 7 from left to right and from right to left, as shown in the lower figure in Figure 7-8. For instance, the determination of the force in the diagonal tie in Panel 2 and the related vertical strut between Panel 2 and 3 are described. The dead load effect is found by summing the bottom numbers under the panel points to the right and subtracting the sum of the top numbers to the left, thus

$$6 + 5 + 4 + 3 + 2 + 1 - 1 = 20$$

This number, divided by the number of panels and multiplied by the panel dead load, gives the vertical component of the strut/tie force due to dead load (14.0 tons). The live load effect is found by summing the bottom numbers only (6 + 5 + 4 + 3 + 2 + 1), giving 21/8 × 10.7 = 28.1 tons. Dead + live load is 38.1 tons, the index written in this panel. The vertical component is equal to the strut compressive force and, multiplied by the ratio 1.32, gives the tie force.

The vertical component of the force in the counter in Panel 4 is found in the same way, by loading panels 1, 2, and 3 to the left of the counter, thus

$$\text{dead load } [1+2+3-(4+3+2+1)] \div 8 \times 5.61 = -2.80$$

$$\text{live load } (1+2+3) \div 8 \times 10.3 = 7.72$$

$$\text{force in counter} = (7.7-2.8) \times 1.32 = 6.5$$

The vertical component of the force in a counter in Panel 3 is

$$\text{dead load } [1+2-(5+4+3+2+1)] \div 8 \times 5.61 = -8.41$$

$$\text{live load } (1+2) \div 8 \times 10.3 = 3.87$$

as the dead load effect in Panel 3 is greater than the live load effect, no counter is required. A check of the table in Figure 7-10 shows that the indices in this table are the same as those determined here.

This force corresponds to a panel shear of 130 tons as shown in Figure 7-8. To find, say, the bottom chord force in the third panel,

$$M = 1/2(2 \times 6)16.3 \text{ tons} \times 22.46 \text{ ft} \div 26 \text{ ft} = 84 \text{ tons}$$

corresponding to a panel shear of 97.2 tons. Finding forces in the web members requires either dividing into dead load and live load forces, or adding a correction force for the missing live loads at some of the panels. The former procedure is illustrated in Box 7-1.

For longer trusses, it generally was desired to maintain an angle of approximately 45 degrees for the braces, but the greater depth of the truss resulted in excessive spans for the floor system. There were a number of remedies for this condition, but the most widespread was the use of a double system of web diagonals, such as shown in Figures 7-11 and 7-12. This system could also take the form of a double intersection truss, or as a truss with subdivided panels (both shown in the boxed material) or in a number of other truss forms, such as the Post truss, the Baltimore truss, or the triangulated Warren truss shown in Figure 7-11. A more complete discussion of index analysis for these truss forms is also given in Swain (1927).

A truss with a double system, such as a Whipple truss (or double-intersection Pratt truss, Figure 7-12) is analyzed as two separate systems. The systems can be represented as solid lines and dashed lines. For instance, for a 16-panel Whipple truss with dead loads of 10,000 at each upper chord panel point and 22,000 at each lower chord, the two systems have the indices shown in Figure 7-12 for the web system.

The index force on the inclined end post, that is, the vertical component of the force is 112 (solid system) + 96 (dashed system) + 22 (bottom chord panel loading) + 10 (top

Figure 7-9. Example of eight-panel Pratt truss road bridge, Waterford, Loudon County, VA (HAER VA,20-CLARK.V,1–5), constructed in 1889.
Source: Photograph by James DuSel.

chord panel loading) = 240 kips. The actual force in the end post is this force times the end post length/panel length.

The remainder of the top chord force can be found by adding in the horizontal force-producing components of the dashed system and solid system loads. The top chord index force in the first panel is 240 + 112 + 2 × 96 kips = 544 kips, so that the force in the top chord in this panel is 566 kips × p/h, where p is the panel length and h is the height of the truss. The top chord force in the middle panel is 566 kips + 2(91 + 64 + 59 + 32 + 27) = 1,112 kips. If the panel aspect ratio is $p/h = 0.5$, then the actual force in the top chord is 534,000. Bottom chord forces may be found similarly and are given below the figure.

Live load forces in the web system are then managed using the subscripted indices used to count the panels from each side. To produce maximum live load force, the panels are loaded up to the panel under consideration. For instance, for the double-intersection Pratt truss shown, given a panel live load of 40 kips, the maximum live load force in the fifth panel is the sum of the lower indices in the panels to the right in the dashed system, that is (11 + 9 + 7 + 5 + 3 + 1)/16 × the panel live load. The index force is 36/16 × 40t = 90 kips.

TABLE XLI.

STRESSES IN SINGLE-INTERSECTION TRUSSES.

w = panel live load on one truss,
W_1 = panel dead load on one truss,
W' = upper panel dead load on one truss,
θ = inclination of diagonal to vertical.

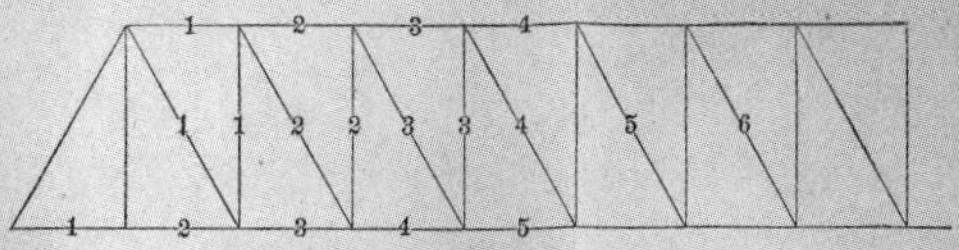

MEMBER.	3 Panel.		4 Panel.		5 Panel.		6 Panel.		7 Panel.		8 Panel.		9 Panel.		Multiply by
	w	W_1	w	W_1	w	W_1	w	W_1	w	W_1	w	W_1	w	W_1	
Top Chord 1	1	1	2	2	3	3	4	4	5	5	6	6	7	7	tan θ
" " 2					3	3	$4\frac{1}{2}$	$4\frac{1}{2}$	6	6	$7\frac{1}{2}$	$7\frac{1}{2}$	9	9	
" " 3									6	6	8	8	10	10	
" " 4													10	10	
Bottom Chord . . . 1	1	1	$1\frac{1}{2}$	$1\frac{1}{2}$	2	2	$2\frac{1}{2}$	$2\frac{1}{2}$	3	3	$3\frac{1}{2}$	$3\frac{1}{2}$	4	4	
" " . . . 2	1	1	$1\frac{1}{2}$	$1\frac{1}{2}$	2	2	$2\frac{1}{2}$	$2\frac{1}{2}$	3	3	$3\frac{1}{2}$	$3\frac{1}{2}$	4	4	
" " . . . 3					3	3	4	4	5	5	6	6	7	7	
" " . . . 4									6	6	$7\frac{1}{2}$	$7\frac{1}{2}$	9	9	
" " . . . 5													10	10	
Batter Brace	$\frac{3}{3}$	1	$\frac{6}{4}$	$1\frac{1}{2}$	$\frac{10}{5}$	2	$\frac{15}{6}$	$2\frac{1}{2}$	$\frac{21}{7}$	3	$\frac{28}{8}$	$3\frac{1}{2}$	$\frac{36}{9}$	4	sec θ.
Diagonal 1	$\frac{1}{3}$	0	$\frac{3}{4}$	$\frac{1}{2}$	$\frac{6}{5}$	1	$\frac{10}{6}$	$1\frac{1}{2}$	$\frac{15}{7}$	2	$\frac{21}{8}$	$2\frac{1}{2}$	$\frac{28}{9}$	3	
" 2			$\frac{1}{4}$	$-\frac{1}{2}$	$\frac{3}{5}$	0	$\frac{6}{6}$	$\frac{1}{2}$	$\frac{10}{7}$	1	$\frac{15}{8}$	$1\frac{1}{2}$	$\frac{21}{9}$	2	
" 3					$\frac{1}{5}$	-1	$\frac{3}{6}$	$-\frac{1}{2}$	$\frac{6}{7}$	0	$\frac{10}{8}$	$\frac{1}{2}$	$\frac{15}{9}$	1	
" 4									$\frac{3}{7}$	-1	$\frac{6}{8}$	$-\frac{1}{2}$	$\frac{10}{9}$	0	
" 5											$\frac{3}{8}$	$-1\frac{1}{2}$	$\frac{6}{9}$	-1	
" 6													$\frac{3}{9}$	-2	
Post (Through-Bridge) 1			$\frac{1}{4}$	$-\frac{1}{2}$	$\frac{3}{5}$	0	$\frac{6}{6}$	$\frac{1}{2}$	$\frac{10}{7}$	1	$\frac{15}{8}$	$1\frac{1}{2}$	$\frac{21}{9}$	2	To the stress on each post must be added W'.
" " " . 2							$\frac{3}{6}$	$-\frac{1}{2}$	$\frac{6}{7}$	0	$\frac{10}{8}$	$\frac{1}{2}$	$\frac{15}{9}$	1	
" " " . 3											$\frac{6}{8}$	$-\frac{1}{2}$	$\frac{10}{9}$	0	
Post (Deck-Bridge) . 1			$\frac{5}{4}$	$-\frac{1}{2}$	$\frac{8}{5}$	0	$\frac{12}{6}$	$\frac{1}{2}$	$\frac{17}{7}$	1	$\frac{23}{8}$	$1\frac{1}{2}$	$\frac{30}{9}$	2	
" " " . 2							$\frac{9}{6}$	$-\frac{1}{2}$	$\frac{13}{7}$	0	$\frac{18}{8}$	$\frac{1}{2}$	$\frac{24}{9}$	1	
" " " . 3											$\frac{14}{8}$	$-\frac{1}{2}$	$\frac{19}{9}$	0	

Figure 7-10. Table showing calculation of indices for single intersection Pratt trusses.
Source: Waddell (1894).

The live load force in the diagonal is then 90 kips $\sqrt{2}$, and the live load force in the post supporting the top of this diagonal is 90 kips. Alternatively, the web system forces may be computed using the full live load, and the sum of the upper indices to the left of the fifth panel may be computed, giving the increase due to the removal of the live load forces from the panel points to the left. The application of the full live load to the dashed system panel points gives a panel index of 80 kips. The increase in index force due to the removal of live load from the first five panels (dashed system only) is $(1 + 3)/16 \times 40$ tons = 10 kips. So the total index force in the fifth panel due to live load is again computed as 90 kips.

Live loads due to "locomotive excess" are also reckoned for the chords and for the web system. The locomotive excess is the amount by which the weight of the locomotive drivers exceeds the distributed train load. It is calculated in an empirical way as the simple difference between the weight of the drivers of one or two locomotives and the following distributed train load. The form of the locomotive excess is one or two concentrated loads

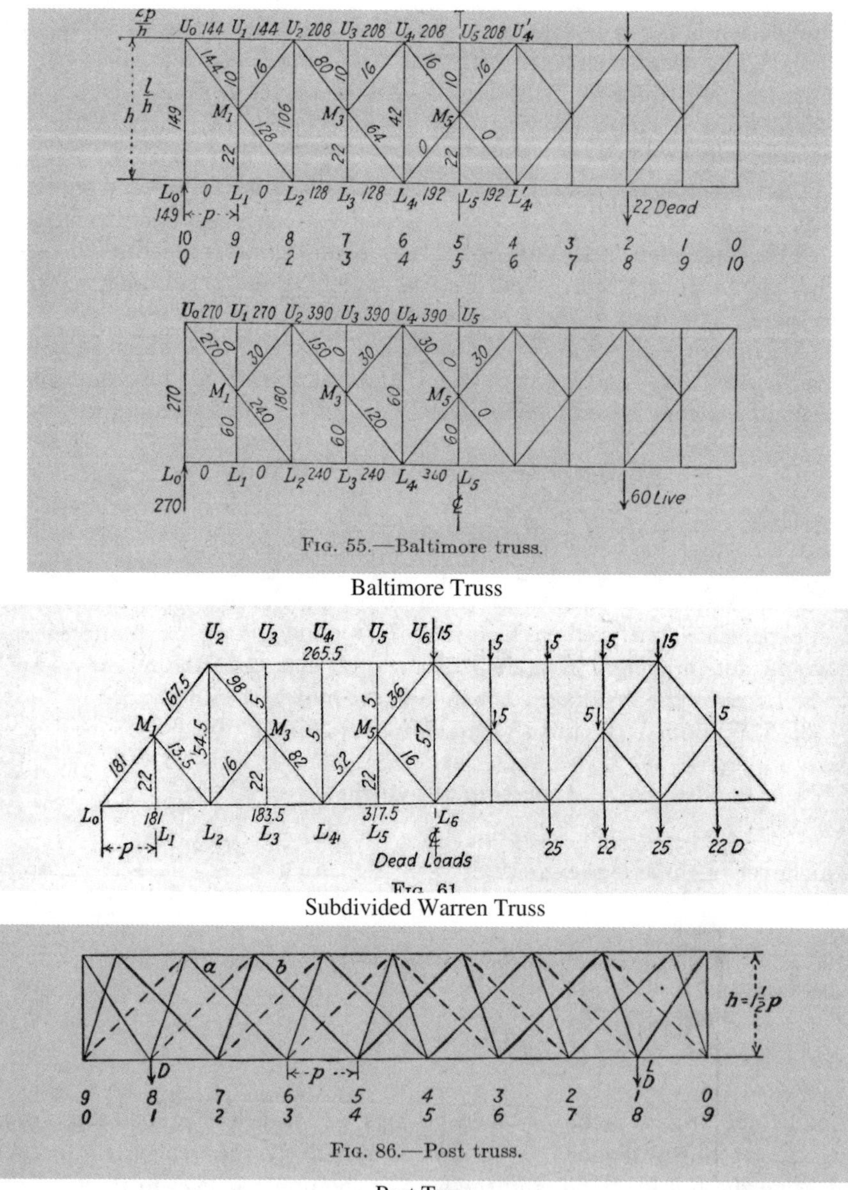

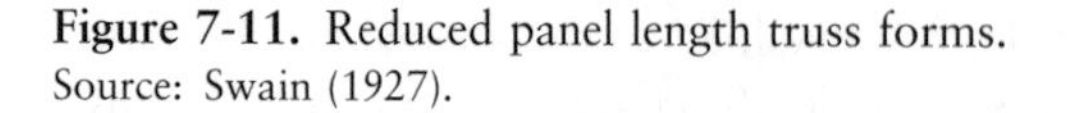
Figure 7-11. Reduced panel length truss forms.
Source: Swain (1927).

at a spacing of 10 to 30 ft. More precise loading conditions were generally handled using Cooper's loading or some other train loading. The handling of locomotive excess loading is described thoroughly in Swain (1896, 1927).

In addition to the described procedure, bar forces in a double-intersection Pratt truss (with 15 or fewer panels) could be determined by consulting a table, such as the one shown

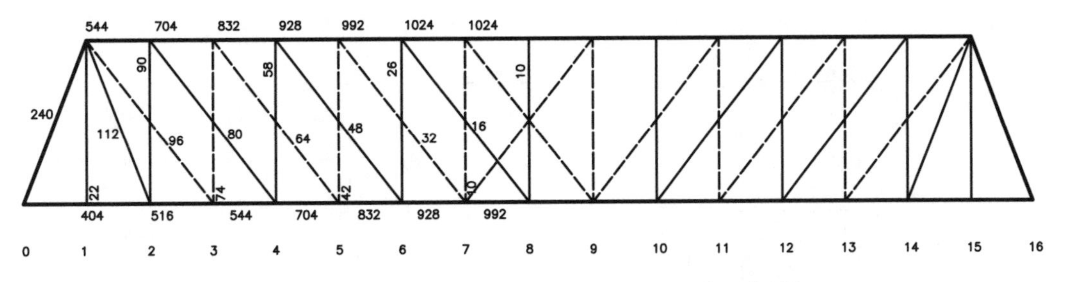

Figure 7-12. Indexing analysis of a double intersection Pratt (or Whipple) Truss.

in Figure 7-13. Each of the bars in the truss is shown in terms of the number of panel loads carried by the bar, whereas the actual force in the bar can be determined by multiplying by the ratio *p*/*h* or *l*/*h*, as appropriate. Indexing can be extended to many other truss forms, including triple and quadruple intersection trusses, Baltimore trusses, Warren trusses, Warren trusses with secondary system, Bollman trusses, Fink trusses, or others. The combination of published tables and the facility with which this method can be extended to trusses not included in the range of published tables make the indexing method a robust procedure for the analysis and design of parallel chord road and rail bridge trusses.

A similar principle may be applied to building trusses in common configurations. The records of the Berlin Iron Bridge Company include a notebook of standard designs for Polonceau or Fink and fan trusses with spans from 30 to 100 ft. These configurations, which are widely used in industrial construction, are properly termed "Polonceau" trusses, after their original author Camille Polonceau (Holzer 2010). However, due to the resemblance of the load paths between this truss and the Fink truss, described previously, this truss form is usually referred to as a "Fink" truss in the late nineteenth century literature. The designs for the 16-panel trusses show evidence of using similar principles to note quickly the forces in the members due to a uniformly distributed load. The task of analyzing these trusses is simplified by standardizing the number of panels (a Fink truss has 4, 8, 16, or 32 panels) and by maintaining a 6:12 roof pitch in all of the examples. In the example of the 90-ft span truss shown in Figure 7-14, with 16 panels of 5.625 ft each, the panel load of 1.26 tons can be resolved into a component parallel to the top chord (0.56 tons) and perpendicular to the top chord (1.12 tons). The bar force of 1.12 tons is first noted on all of the odd numbered panels, where the purlin load is the only load acting on the truss bars. This force can be resolved into a force along the adjacent web members and recognized as equal in magnitude to the panel load, so the force in each of these elements is noted as 1.26 tons. The force in the web members incident at even numbered panels, is the sum of one panel load plus two half-panel loads transmitted from the adjacent panels, hence 2.25, whereas the force in the long web member is double this: 4.50. All of the bars perpendicular to the top chord are in the same geometric relationship, so the forces in these bars are 1.12, 2.25, or 4.50. The remainder of the web member forces can be completed on similar principles. The bottom chord forces can be found by finding the horizontal component of the reaction and subtracting 1.26 tons at the first joint, 2.52 at the second, and 5.04 at the third. The top chord force can be found by determining the force of 21.15 at the support and subtracting 0.56 tons at each successive panel point.

TABLE XLII.

STRESSES IN DOUBLE-INTERSECTION TRUSSES.

w = panel live load on one truss,
W_1 = panel dead load on one truss,
W' = upper panel dead load on one truss,
α = inclination of short diagonal to vertical,
β = inclination of long diagonal to vertical.

MEMBER.	7 Panel.		8 Panel.		9 Panel.		10 Panel.		11 Panel.		12 Panel.		13 Panel.		14 Panel.		15 Panel.		Multiply by
	w	W_1	w	W_1	w	W_1	w	W_1	w	W_1	w	W_1	w	W_1	w	W_1	w	W_1	
Top Chord 1	40/7	40/7	7	7	74/9	74/9	9½	9½	118/11	118/11	12	12	172/13	172/13	14½	14½	236/15	236/15	tan α.
" " 2	44/7	44/7	8	8	88/9	88/9	11½	11½	146/11	146/11	15	15	218/13	218/13	18½	18½	304/15	304/15	
" " 3	44/7	44/7	8	8	92/9	92/9	12½	12½	162/11	162/11	17	17	250/13	250/13	21½	21½	356/15	356/15	
" " 4					92/9	92/9	12½	12½	168/11	168/11	18	18	270/13	270/13	23½	23½	394/15	394/15	
" " 5									168/11	168/11	18	18	276/13	276/13	24½	24½	416/15	416/15	
" " 6													276/13	276/13	24½	24½	424/15	424/15	
" " 7																	424/15	424/15	
Bottom Chord . . . 1	21/7	21/7	3½	3½	30/9	30/9	4½	4½	55/11	55/11	5½	5½	78/13	78/13	6½	6½	105/15	105/15	
" " . . . 2	21/7	21/7	3½	3½	30/9	30/9	4½	4½	55/11	55/11	5½	5½	78/13	78/13	6½	6½	105/15	105/15	
" " . . . 3	30/7	30/7	5	5	52/9	52/9	6½	6½	80/11	80/11	8	8	114/13	114/13	9½	9½	154/15	154/15	
" " . . . 4	36/7	36/7	7	7	74/9	74/9	9½	9½	118/11	118/11	12	12	172/13	172/13	14½	14½	236/15	236/15	
" " . . . 5					84/9	84/9	11½	11½	146/11	146/11	15	15	218/13	218/13	18½	18½	304/15	304/15	
" " . . . 6									156/11	156/11	17	17	250/13	250/13	21½	21½	356/15	356/15	
" " . . . 7													264/13	264/13	23½	23½	394/15	394/15	
" " . . . 8																	408/15	408/15	
Batter Brace	21/7	21/7	28/8	3½	30/9	30/9	45/10	4½	55/11	55/11	66/12	5½	78/13	78/13	91/14	6½	105/15	105/15	sec α.
Diagonal 1	9/7	9/7	12/8	1½	16/9	16/9	20/10	2	25/11	24/11	30/12	2½	36/13	36/13	42/14	3	48/15	48/15	
" 2	6/7	5/7	9/8	1	12/9	11/9	16/10	1½	20/11	19/11	25/12	2	30/13	28/13	35/14	2½	42/15	41/15	sec β.
" 3	4/7	2/7	6/8	½	8/9	7/9	12/10	1	16/11	14/11	20/12	1½	25/13	23/13	30/14	2	36/15	34/15	
" 4	3/7	−2/7	4/8	0	6/9	3/9	9/10	½	12/11	8/11	16/12	1	20/13	17/13	25/14	1½	30/15	26/15	
" 5			2/8	−½	4/9	−1/9	6/10	0	9/11	3/11	13/12	½	16/13	10/13	20/14	1	25/15	18/15	
" 6					2/9	−5/9	4/10	−½	6/11	−3/11	9/12	0	13/13	3/13	16/14	½	20/15	11/15	
" 7							2/10	−1	4/11	−8/11	6/12	−½	9/13	−3/13	12/14	0	16/15	4/15	
" 8											4/12	−1	6/13	−10/13	9/14	−½	12/15	−4/15	
" 9													4/13	−16/13	6/14	−1	9/15	−11/15	
" 10																	6/15	−18/15	
[illegible] 1	4/7	2/7	6/8	½	8/9	7/9	12/10	1	16/11	14/11	20/12	1½	25/13	23/13	30/14	2	36/15	34/15	
[illegible] 2			4/8	0	6/9	3/9	9/10	½	12/11	8/11	16/12	1	20/13	17/13	25/14	1½	30/15	26/15	
" " " 3			2/8	−½	4/9	−1/9	6/10	0	9/11	3/11	13/12	½	16/13	10/13	20/14	1	25/15	18/15	
" " " 4							4/10	−½	6/11	−3/11	9/12	0	13/13	3/13	16/14	½	20/15	11/15	
" " " 5											6/12	−½	9/13	−3/13	12/14	0	16/15	4/15	
" " " 6															9/14	−½	12/15	−4/15	
Post (Deck Bridge) . 1	11/7	2/7	14/8	½	18/9	7/9	22/10	1	27/11	14/11	32/12	1½	38/13	23/13	44/14	2	51/15	34/15	To the stress on each post must be added W'.
" " " . 2	2/7	−2/7	12/8	0	15/9	3/9	18/10	½	23/11	8/11	28/12	1	33/13	17/13	39/14	1½	45/15	26/15	
" " " . 3			10/8	−½	13/9	−1/9	16/10	0	20/11	3/11	24/12	½	29/13	10/13	34/14	1	40/15	18/15	
" " " . 4							14/10	−½	17/11	−3/11	21/12	0	25/13	3/13	30/14	½	35/15	11/15	
" " " . 5											18/12	−½	22/13	−3/13	26/14	0	31/15	4/15	
" " " . 6															23/14	−½	27/15	−4/15	

Figure 7-13. Tabulated index stresses for a double intersection Pratt truss.
Source: Waddell (1894).

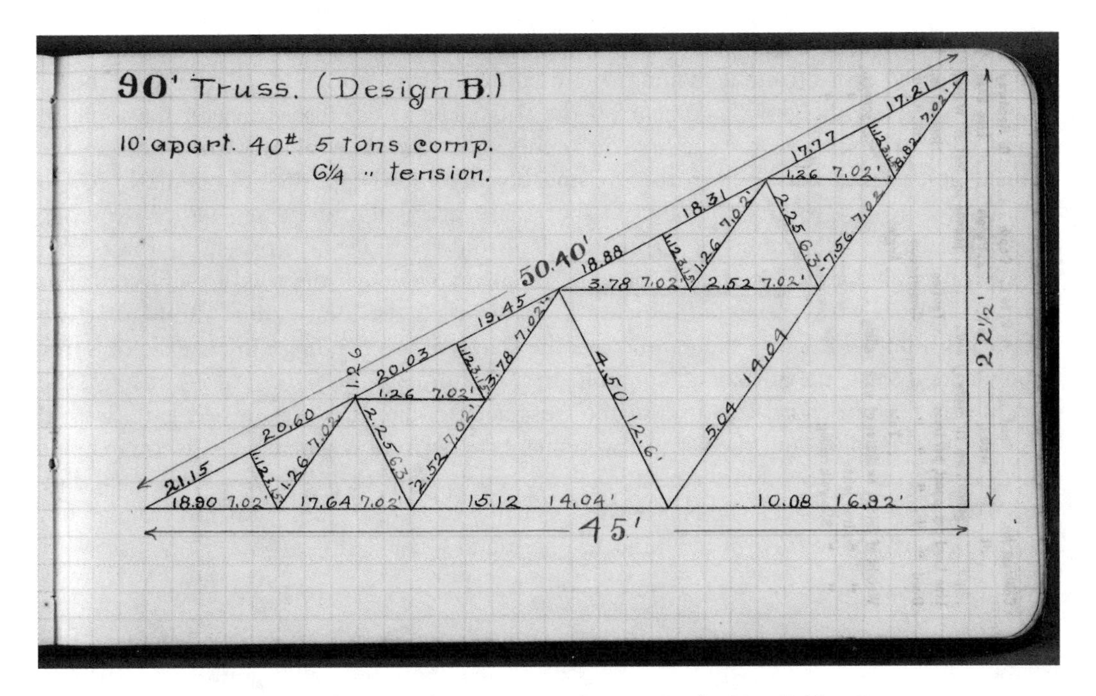

Figure 7-14. Analysis of 90-ft span Polonceau or Fink truss, Berlin Iron Bridge Company.
Source: Reproduced by permission from The Huntington Library, San Marino, CA.

Analysis of Curved Chord Bridges

The principal authors describing the methods of analysis for curved chord bridges are Swain (1896) and Du Bois (1888). Du Bois's methods rely in general on the method of sections (also known as method of moments or Ritter's method) for the determination of the maximum forces in the bars of the truss. Although Swain uses the method of sections, he also determines algebraically some surprisingly simple procedures for the determination of member forces, at least for various forms of parabolic trusses. Finally, it is possible to show to what extent some of the procedures were used in practice by examination of the design standards of the Berlin Iron Bridge Company, which was a leading late nineteenth-century manufacturer of double parabolic, or lenticular, trusses.

According to Du Bois (4th ed., 1888, p. 123), the readiest method of solution for such bridges is the graphical method, at least for the fully loaded truss, whereas the readiest analytical method is the method of moments. Although Du Bois proposes using the graphical method as a check on analytical methods for highway and rail bridges with their variable loading, he and other authors consider the graphical method too cumbersome for everyday use. He makes the calculation of the height of the truss and the lever arms to use for the method of moments an initial stage of the calculation of the forces in the truss. The forces in the braces, that is, the web members, have to be calculated separately under individual sets of loads. Thus, the calculation of curved chord trusses is always a methodical procedure: find the geometry and lever arms for use in force calculation, find chord forces under full loading, and find brace and post forces under partial loading, all by the method of moments.

Swain (1896) uses similar arguments to develop the analysis of parabolic and other forms of curved chord truss. Beginning with a parabolic top chord and a flat bottom chord, Swain investigates the global moment on the truss. His initial discussion on the topic of curved chord trusses uses the method of moments for the determination of forces in the inclined ties, in the posts, and in the top and bottom chord (pp. 46–50), similarly to Du Bois. Following this, Swain enters into a specialized discussion of trusses with a parabolic chord. The basis of this discussion is the fundamental relationship between the forces in a truss with parabolic chords and the bending moments to which the truss is subjected. For a parabolic truss with a level bottom chord, for instance, the forces under full loading are particularly simple. Because the parabola is the shape of the bending moment diagram for a fully loaded truss, the horizontal component of the force in the top chord is constant; the force in the bottom chord is constant; the truss acts fundamentally as a tied arch, with 0 force in the diagonals; and the vertical web members act simply as hangers. Swain extends this discussion to trusses with a parabolic top and bottom chord in any form, including lenticular trusses and sickle-shaped trusses. On this basis, the horizontal force under full loading in a parabolic truss of any kind can be found a constant:

$$H = wl^2/8h_c$$

where

H is the horizontal component of the force in any top or bottom chord bar;
w is the uniformly distributed live load (Swain uses p, which is easily confused with the panel length);
h_c is the height of the truss at the center; and
l is the span of the truss.

This can be put in a form more consistent with the later discussion, in which the panel length is factored out of most of the equations

where
H is $(wa)am^2/8h_c$,
a is the length of a panel (wa = the panel load P), and
m is the total number of panels (am = the span):

$$H = Pal/8h_c$$

The analysis of the web members yields a more surprising result. The web, as described, is neutral under full loading of the truss, with the verticals acting as hangers. To find the maximum force in a web diagonal, it is necessary to place the live load only on each of the panel points between the diagonal and the abutment in the manner shown in Figure 7-15, where the loading is place to create the maximum force in the diagonal in the third panel from the left (shown bold). For this loading, the force can be obtained by a general method of sections. The result in Figure 7-16 shows that the horizontal component of the maximum diagonal force is equal to the differences in the coefficients M_1/h_1 and M_2/h_2 on either end of the diagonal. For this form of variable loading, the horizontal component of the force in

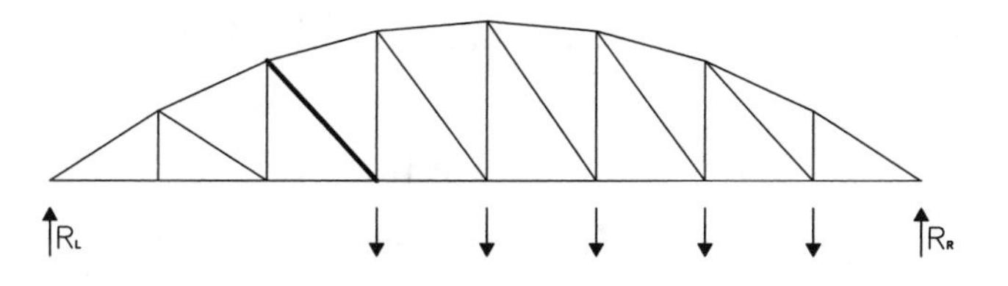

Figure 7-15. Live load placement for maximum stress in a diagonal.

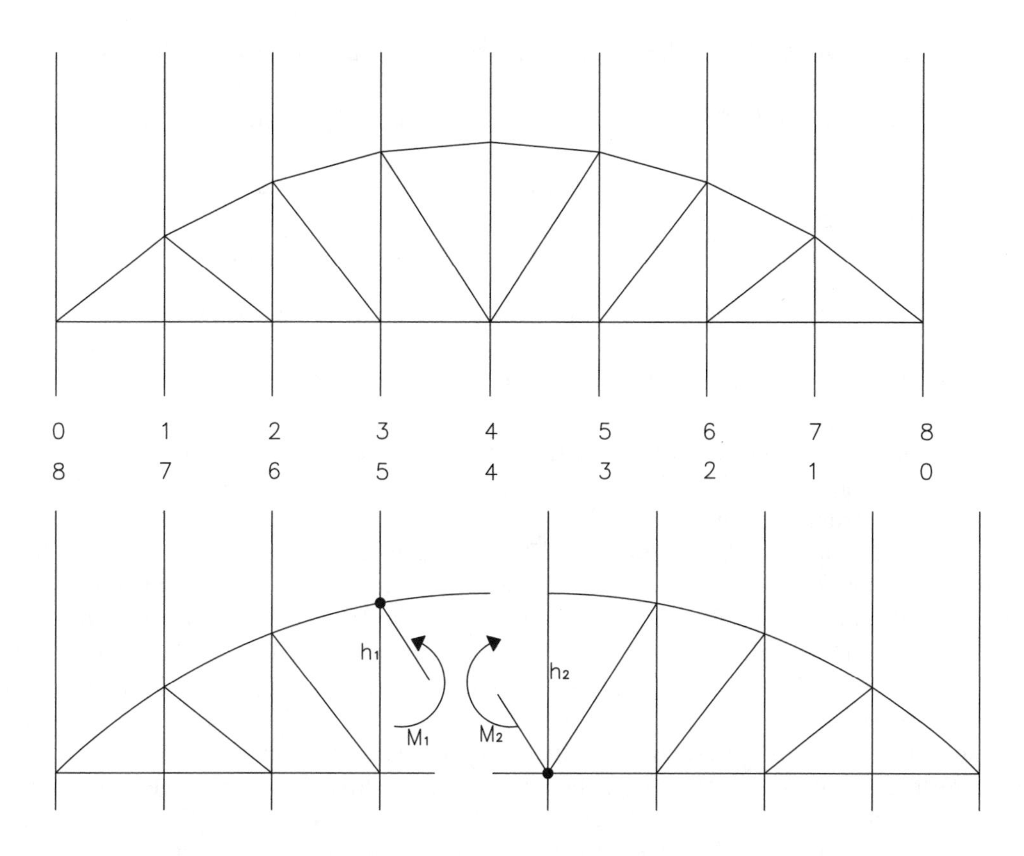

Horizontal component bottom chord = M_1/h_1
Horizontal component top chord = M_2/h_2
Horizontal component in diagonal = $M_1/h_1 - M_2/h_2$

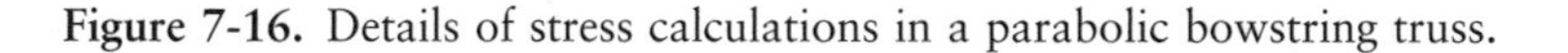

Figure 7-16. Details of stress calculations in a parabolic bowstring truss.

a web diagonal in a parabolic truss is constant. In the end, the analysis of a parabolic truss requires the calculation of two quantities: the horizontal component of the chord force and the horizontal component of the force in the web diagonals. For this truss, or any truss, the horizontal component of force in the bottom chord is equal to M_1/h_1, the horizontal force in the top chord is M_2/h_2, and the horizontal component of the force in the diagonal is the difference between the two. In the truss illustrated in Figure 7-16, the top chord horizontal component, say, in Panel 3 is

$$M_2/h_2 = 1/2(3\times 5)Pa \div [(3\times 5)/16]h_4$$
$$= (8Pa/h_4)\ (a = \text{panel length};\ P = \text{panel load})$$

Whereas the bottom chord horizontal component in the same panel is

$$M_1/h_1 = 1/2(2\times 6)Pa \div (2\times 6/16)h_4 = (8Pa/h_4)$$

For the diagonal in Panel 3 under the loading causing maximum force,

$$M_1/h_1 = (1/8)(5+4+3+2+1)P\times 2a \div [(6\times 2)/16]h_4 = 5.0\ Pa/h_4$$

$$M_2/h_2 = (1/8)(5+4+3+2+1)P\times 3a \div [(5\times 3)/16]h_4 = 6.0\ Pa/h_4$$

For the diagonal in Panel 4 under the loading causing maximum force,

$$M_1/h_1 = (1/8)(4+3+2+1)P\times 3a \div [(5\times 3)/16]h_4 = 4.0\ Pa/h_4$$

$$M_2/h_2 = (1/8)(4+3+2+1)P\times 4a \div [(4\times 4)/16]h_4 = 5.0\ Pa/h_4$$

so the horizontal component of force in the diagonal can be seen to be equal between the two panels or more generally among all the panels. A complete calculation of the bar forces in a parabolic truss is given in Box 7-2.

A contemporary example (Figure 7-17) shows the calculations from the Berlin Iron Bridge Company from about 1894 on a parabolic lenticular truss bridge, 10 panels in length, with a total span of 150 ft and a center height of 22 ft. Panel loads are 7.5 tons live and 3.0 tons dead. The depth of the truss is 27 ft at the center. The calculated horizontal force in the chords is 72.91 tons, resulting in a total maximum chord force of 76.64 in the panel adjacent to the support. This can be compared to the force of 76.50 shown on the drawing. For this truss, it is not necessary to calculate the force in the remaining chord elements, because it is not proposed to change the chord section along its length. The force in the center vertical, calculated by the aforementioned formula is +1.50 dead, 7.5 live (one full panel live load), −6.00 total, whereas the horizontal component of the maximum force in a

Box 7-2

Swain (1927) calculates all of the bar forces in a parabolic truss with horizontal bottom chord, and in a lenticular truss, with parabolic bottom chord, using the integers representing panel numbers and the parameters of the center height of the truss h_c, the panel length a, the distributed live load w, and the distributed dead load g.

Swain begins by expressing the height of the bridge at any point x:

$$h = h_c(4x/l)(1 - x/l)$$

As shown in Figure B7-2-1, we want to find the maximum force in the diagonal n_1, or at $x = n_1 a$ (shown in bold). So there are $m - n_1$ loaded panels, where m is the total number of panels in the bridge. Then, the left-hand reaction R_L can be found to be

$$R_L = wa(m - n_1)\frac{(m - n_1 + 1)}{2m}$$

from which

$$M_1 = wa^2 n_1 (m - n_1)\frac{(m - n_1 + 1)}{2m}$$

$$M_2 = wa^2 (n_1 - 1)\left[(m - n_1)\frac{(m - n_1 + 1)}{2m}\right]$$

Now, by the formula given for the height of the truss at a point

$$h_1 = \frac{4h_c n_1}{m}\left(1 - \frac{n_1}{m}\right)$$

and

$$h_2 = \frac{4h_c (n_1 - 1)}{m}\left[1 - \frac{(n_1 - 1)}{m}\right]$$

M_1/h_1 and M_2/h_2, per the previous discussion, can be used to compute the horizontal component of the stress in the diagonal:

$$\frac{M_1}{h_1} = wa^2 m\frac{(m - n_1 + 1)}{8h_c}$$

$$\frac{M_2}{h_2} = wa^2 m\frac{(m - n_1)}{8h_c}$$

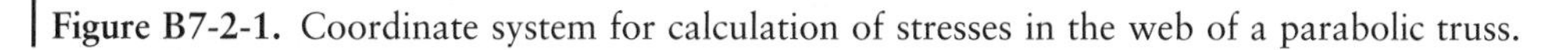

Figure B7-2-1. Coordinate system for calculation of stresses in the web of a parabolic truss.

As a result, the horizontal component of stress in the diagonal $M_1/h_1 - M_2/h_2$ is found to be equal to $wa^2m/8h_c$, that is, independent of n, which is constant throughout the truss.

For a parabolic lenticular truss (illustrated in Figure B7-2-2), the forces in the top and bottom chord, and in the diagonals, can be found from these equations without modification. Although the dead load force in a vertical is simply half of the panel load, the maximum live load force in a vertical is found by taking a section, such as the one shown in Figure B7-2-2, for the loading shown. As a result, the total compressive live load force is equal to the reaction less the vertical component of the force in the top chord bar to the left, less the vertical component of the bottom chord force in the bar to the right. Swain (1927) shows that this force again reduces to a particularly simple formula.

Starting from the same expression for the R_L as used in the discussion, he determines the horizontal force component in the upper or lower chord in the section under investigation as M_1/h_1, as

$$HC = wa^2 \frac{(m - n_1 + 1)m}{8h_c}$$

Based on the geometry of the upper chord, the vertical component of the upper chord (to the left) can be found to be

$$VCU = \frac{wa(m - n_1 + 1)(m - 2n_1 + 1)}{4m}$$

Similarly, the vertical component in the lower chord (to the right) is

$$VCL = \frac{wa(m - n_1 + 1)(m - 2n_1 - 1)}{4m}$$

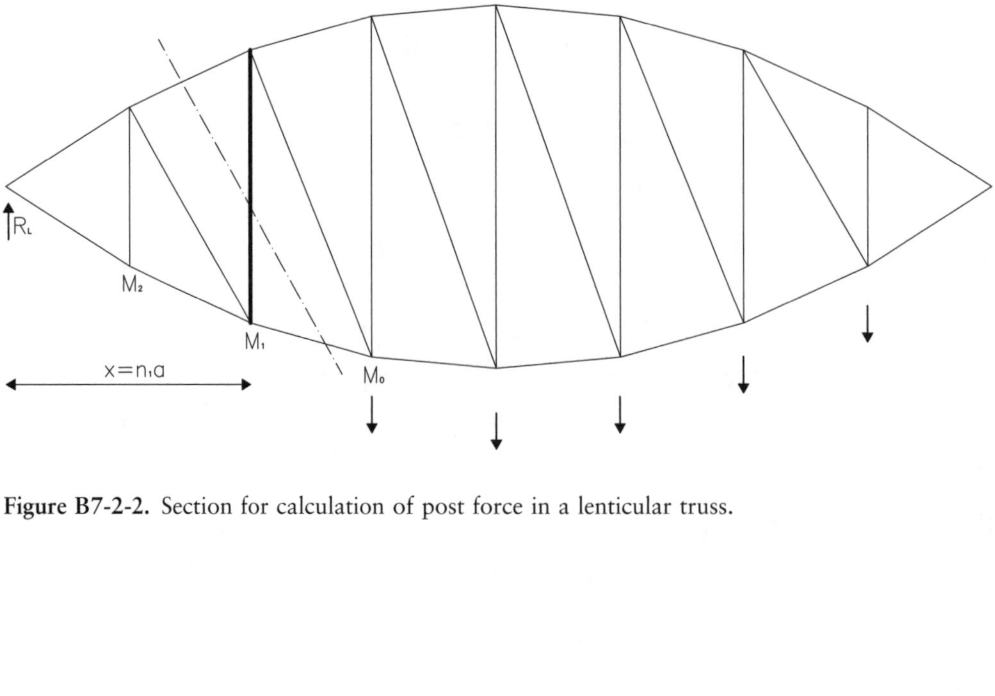

Figure B7-2-2. Section for calculation of post force in a lenticular truss.

Finally, subtracting VCU and VCL from R_l gives the force V in the vertical

$$V = wan_1 \frac{(m - n_1 - 1)}{2m}$$

The remaining calculations, which involve finding vertical components corresponding to the horizontal components in the chords and in the diagonals, are facilitated by developing similar formulas for the ratio of the vertical component to the horizontal component, based on the geometry of the truss. Such formulas do not appear in Swain's work, but they are presented here.

For the chord, for a truss with an even number of panels the ratio of vertical/horizontal component for panel n_1 is

$$\frac{VC}{HC} = (m - 2n_1 + 1)\frac{4h_c}{am^2}$$

whereas this ratio for the web diagonals is

$$\frac{VC}{HC} = [(2n_1 - 1)(m - n_1) + (n_1 - 1)]\frac{2h_c}{am^2}$$

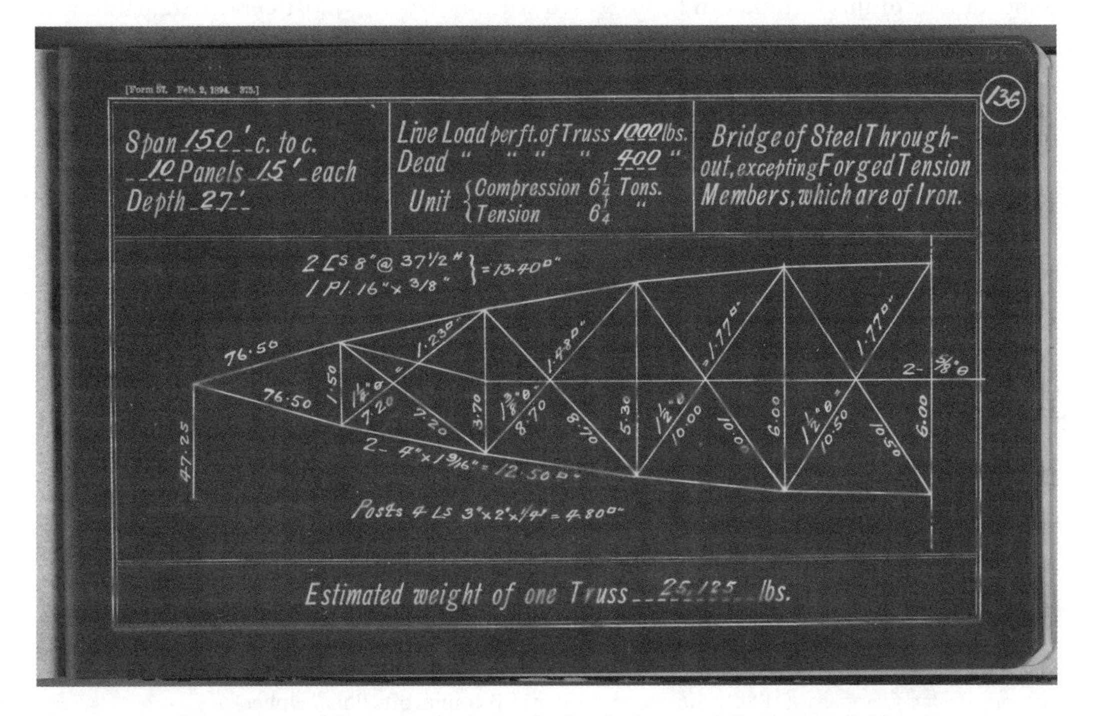

Figure 7-17. Calculations of forces in the bars of a lenticular truss, Berlin Iron Bridge Company. Source: Reproduced by permission from The Huntington Library, San Marino, CA.

TABLE 7-1. *Calculation of Bar Forces in a Lenticular Truss*

Panel	Diagonal *HC*	Diagonal *VC/HC*	Diagonal Resultant Calculated	Diagonal Resultant From Drawing	Vertical Dead	Vertical Live	Vertical Total Calculated	Vertical Total From Drawing
1	5.21				+1.50	−3.00	−1.50	1.50
2	5.21	0.900	7.00	7.20	+1.50	−5.25	−3.75	3.70
3	5.21	1.58	8.68	8.70	+1.50	−6.75	−5.25	5.30
4	5.21	1.62	9.92	10.00	+1.50	−7.50	−6.00	6.00
5	5.21	1.76	10.56	10.50	+1.50	−7.50	−6.00	6.00

diagonal is 5.21 tons, resulting, for instance, in a total force of 10.56 tons. A full calculation of the forces in the web members is shown in Table 7-1. All values are tabulated in tons.

Various analytical methods were available to late nineteenth-century bridge designers for the calculation of forces in bridge trusses under variable live loading. These methods tended to be specialized to one form of truss or another. The indexing method generally is suitable for parallel chord trusses; although complications arise in more intricate bridge forms, such as the Baltimore truss, the method is well adapted for rapid analysis of the single- and double-intersection Pratt, Warren, and Howe trusses. A general method, the method of moments, can be practiced for bridges with curved chords, and the laborious computations of this method can be dispensed with for trusses whose chords are parabolic. The most widely used analytical computation methods appear to be the indexing method and the direct method for calculating the forces in trusses with parabolic chords.

References Cited

Bow, R. H. (1874). *A treatise on bracing*. Van Nostrand, New York (originally published in Britain in 1851).

Crandall, C. L. (1888). *Notes on bridge stresses and bridge designing*. Mimeographed lecture notes, Cornell University Civil Engineering Department, Ithaca, NY.

Crehore, J. D. (1886). *Mechanics of the girder*. John Wiley and Sons, New York.

Du Bois, A. J. (1888). *The strains in framed structures*, 4th Ed. John Wiley and Sons, New York.

Dufour, F. O. (1909). *Bridge engineering. Roof trusses*. American School of Correspondence, Chicago.

Hatfield, R. G. (1871). *The American house-carpenter*, 7th Ed. John Wiley and Sons, New York.

Haupt, H. (1856). *General theory of bridge construction*. D. Appleton, New York.

Henrici, O. (1867). *Skeleton structures especially in their application to the building of steel and iron bridges*. Van Nostrand, New York.

Holzer, S. (2010). The Polonceau roof and its analysis. *International Journal for the History of Engineering and Technology*, 80, 22–54.

Merrill, W. (1870). *Iron truss bridges for railroads*. Van Nostrand, New York.

Phoenix Bridge Company. (1888). *Album of designs*. J.B. Lipincott, Philadelphia.

Sheilds, F. W. (1867). *The strains on structures of ironwork*, 2nd Ed. John Weale, London.

Stoney, B. B. (1873). *The theory of strains in girders and similar structures*. Van Nostrand, New York.

Swain, G. F. (1896). *Notes on the theory of structures*, 2nd Ed. Mimeographed lecture notes. Massachusetts Institute of Technology, Department of Civil Engineering, Cambridge, MA.

Swain, G. F. (1927). *Structural engineering, vol. 3. Stresses, graphical statics, and masonry*. McGraw-Hill, New York.

Tredgold, T. (1888). *Elementary principles of carpentry*, 6th Ed. E and F.N. Spon, London.

Waddell, J. A. L. (1894). *The designing of ordinary iron highway bridges*, 5th Ed. John Wiley and Sons, New York.

8

Analysis of Girders: Beams, Plate Girders, and Continuous Girders

Wood, Wrought-Iron, and Cast-Iron Girders

Beams and girders, made of cast iron, plate iron, lattice work, or sawn wood, were in widespread use in the nineteenth century for the support of floors, roofs, and bridge structures. Several methods for proportioning wood girders were practiced: some of the widely used empirical methods are covered in Chapter 3, but methods based on bending theory, either through the development of semiempirical rules of thumb, or through direct analysis, were also used and are discussed in this chapter. Similarly, iron plate and lattice girders were designed both by empirical methods and by analytical methods. Whereas the latter are the primary focus of this chapter, the former are also discussed, as the boundary between empirical and analytical design is indistinct in the design of this type of structure. The development of iron girder design generally follows a pattern from the empirical to the analytical, with many intermediate procedures that rely on empirical ideas. Examples of such ideas are "bending moments in an I-beam are carried by the flanges," or "girder continuity doesn't contribute to strength." By the end of the nineteenth century, design of girders was primarily analytical. The application of continuous girders depended on the production of girders that were long enough for multiple spans and on the development of analytical procedures that were equal to the task of calculating bending moments in these girders.

Strongly held opinions about the merits of continuity in flexural members, especially in long-span steel bridges, were discussed in print. The debate between Mansfield Merriman (1876) and Charles Bender

(1876) is particularly revealing. The proponents of continuity focused on the potential material saving, while opponents argued that such structures were sensitive to settlement and could not be calculated effectively. However, some interesting means of designing continuous girders were widely presented. These are also discussed in this chapter.

Prior to the second half of the nineteenth century, simple rules prevailed, such as those used in London building regulations after the Great Fire of 1666, and those presented in builder's manuals. These rules focus on limiting the span lengths of joists and girders (Yeomans 1987). Thomas Tredgold (1820) and Peter Nicholson (1826) present formulas for the size of wood members, in part based on the experiments of Peter Barlow. Barlow's results were presented in the form of a central breaking load $W = 4ad^2S/L$, where S is a constant determined by experiment, d is the depth of the beam, L is the span length, and a is the width of the beam. These rules, semiempirical in nature, simplified the analysis of wood beams to the point where the rules could be applied by nontechnically trained mechanics. Such recipes also served in cases where the loading is complex, such as a stair trimmer. In cases of multiple loads, Frank Kidder (1886) recommends designing two or more beams of equal depth, assigning the various loads to different beams, and combining the width of the beams. Rule making is used to cover other structures: beams in general, trussed beams, and others by Kidder. Further information on the analysis and design of wood beams is available in Chapter 3.

For both cast-iron and wrought-iron girders, forms of analysis ranging from empirical to rational also were employed. Empirical design was certainly applied to the determination of appropriate span/depth ratios for iron girders. For the stress analysis of the girders themselves, semiempirical rules similar to those used in wood girder design also were widely circulated for iron girder design. There seem to be occasional transitional types of analysis between a truss and a girder: plate bracing is considered a form of bracing in a girder by Bow (1874, pp. 29–30), who deals with the subject qualitatively and by comparison to a lattice girder. Otherwise, emerging ideas about flexural analysis were applied to the analysis of these structures—some methods ignoring the web contribution to moment resistance, others taking account of the web for flexure. The design of the web seems to be particularly full of uncertainties, and various methods for proportioning the web were advanced.

Simply supported girders were by far the most common configuration for bridges and buildings in the late nineteenth century. Continuous girders were used in the form of trusses for which the chords are continuous over the supports. Few instances of continuous girders are available, especially in fixed-span bridges. However, the use of continuity in some form was impossible to avoid in movable bridges. In his course notes (1896, p. 121), Swain asserts, "Continuous girders are never built in this country except for swing bridges .... In Europe, however, continuous girders are often preferred, and French engineers rarely build girders of over one span without making them continuous." Contentious discussions arose in the literature over the merits and demerits of continuous span fixed-span bridges (for example, the exchange between Merriman [1876] and Bender [1876]), but ultimately the profession in the United States appears to have decided in favor of multiple simple span girders for these bridges. The cantilever girder, a reasonable alternative to a statically indeterminate multispan girder, has often been used in place of a continuous span girder in construction. An example of a two-span continuous girder bridge from 1889, seen just prior to its demolition, has been documented by Historic American Engineering Record (HAER) (Figure 8-1). Similarly, an 1890 swing bridge documented by HAER is shown in Figure 8-2. Although,

Figure 8-1. Two-span Memorial Ave. Bridge over Lycoming Creek, Williamsport, PA (demolished), built 1889 (HAER PA,41-WILPO,3–9).
Source: Photograph by Lawrence Mohar.

according to Kidder (1886, p. 327), "girders resting on three or more supports are of quite frequent occurrence in building construction," it is difficult to find examples of intentional use of continuous girders in building applications, and the topic of continuity in building structures is left to the discussions of portal frames in Chapters 10 and 15.

Simply Supported Iron Girders

Simply supported girders generally fell into three classes: cast-iron beams, plate girders, and lattice girders. Lattice girders, however, were analyzed by similar methods to truss analysis, and the discussion of the analysis of trussed girders is found in Chapter 7.

Other authors have their own formulas for computing the dead load of bridges. Isami Hiroi (1893) presents similar formulas for dead loads of rail bridges, and John Alexander Low Waddell (1894) presents extensive tables for the weight of highway bridges. Girders were widely used, and there is a need for the rapid determination of stresses in these

Figure 8-2. Bridgeport Swing Span Bridge, Bridgeport, AL (demolished), built 1890 (HAER ALA,36-BRIPO.V,1-).
Source: Photograph by C. N. Beasley.

elements, whether for rail bridges, road bridges, or buildings. For girders used as floor beams or other repeated elements, the stresses are determined from bending moments calculated for uniformly distributed loads or simple concentrated loads, as appropriate. These beams rarely exceed 20 or 25 ft in span and are made with plate or latticed webs. When the floor beams repeated sufficiently, they were often tapered to reduce overall iron or steel weight.

Various ideas of proportional design were evident in the nineteenth century. As an example of these proportioning rules, Fleeming Jenkin (1873, p. 297) prefers wrought iron for spans of more than 30 ft. Plate girders have spans to 100 ft, lattice girders beyond. According to Jenkin, the depth/span ratio ranges from 1/8 to 1/15 of span. Milo Ketchum (1903, p. 221) has a brief discussion of the design of plate girders. According to him, the flanges are designed for the entire moment, so that the flange force is equal to the bending moment divided by girder depth. Ketchum further states that the web generally can account for one-sixth to one-eighth of the bending moment, based on the amount of perforation. Ketchum's statement about the neglect of web contribution to moment resistance can be verified from several other sources. Kidder (1886, p. 347) also presents a rule for the design of girder flanges, similar to the Ketchum's rule as described, in which the flanges alone are considered to provide bending resistance, and the contribution of the web is neglected. It is reduced to a rule, similar to the rules previously examined. According to this rule, the safe load in tons, taking the allowable stress in the iron as 5 tons/in.^{2} (which includes an allowance for rivet holes) is equal to

$$\frac{10 \times \text{area of one flange} \times \text{height of web in inches}}{3 \times \text{span in feet}}$$

Further assumptions in the development of this formula are that the web does not contribute to bending resistance and that the difference in stress between the extreme fiber of the flange and the location of the flange centroid is offset by neglecting the contribution of the web to bending resistance. This formula can be inverted to find a rule for the sizing of the flange, so the required flange area, in square inches, is

$$\frac{3 \times \text{the total load in tons} \times \text{span in feet}}{10 \times \text{the height of the web in inches}}$$

The application of these textbook and manual methods can be verified from the girder designs for bridges used as office standards by the Berlin Iron Bridge Company. In this book a series of standard designs for steel girder bridges are presented for various loading and support conditions. The bending analysis undertaken for these girders uses the simplification of dividing the bending moment by the height of the web plate to determine the flange force, and the stress in the flange (which consists of double angles riveted to the web plate) is apparently found by dividing the flange force by the net area (less rivet holes). This procedure is not followed by Swain (1896), who considers the computation of the web's resistance to bending along with the flange (p. 7, for instance).

In working with buildings, Kidder (1886, pp. 280–303) provides rules of thumb and load tables for the rolled members available at the time. Sizes of rolled beams up to 15 in. deep appear to be suitable for spans up to 25 ft or more. For larger spans, it is probable that trusses or latticed girders would be used in preference to filled girders in building design. The load tables are presented in the form of safe uniform loads per foot of span, and the tabular value needs to be simply divided by the square of the span for a uniformly distributed load and other coefficients applied for other loading conditions, such as halving the capacity for a concentrated load at mid-span.

Bindon Blood Stoney (1873) recommends using plate iron for the higher shear zones near the supports. William Humber (1869) says that longitudinal strains taken by flange, shear by web (p. 26). An interesting theory of strains in webs of girders is advanced in Stoney's "Chapter Concluding Remarks." Natural strain trajectories are modified by resolution into the actual direction of the members of a latticed web structure. Francis Campin (1868, pp. 26–28) derives the expression $(wx/2D)(L - x)$ for the flange force in a uniformly loaded girder at any point. In this expression w is the uniformly distributed load (lbs/ft), x is the distance from the left support in feet, D is the depth of the girder in feet, and L is the span of the girder in feet. Humber (1869, pp. 26–27) suggests using the web for shear only and calculating the stress in the flange directly by dividing the bending moment by the distance between the flanges, either algebraically or graphically. Where the flanges are curved, Campin adds a correction of the secant of the angle of the tangent to the curve, which is more easily calculated graphically. Jenkin (1873, p. 294) proportions wrought-iron girder flanges similarly, allocating more material to the top flange due to the lesser strength in compression. He uses ultimate strength values of 25 tons/in.^{2} in tension and 20 tons/in.^{2} compression—with a safety factor of 4; these become the working stresses used by a number of authors, including Waddell (1894, p. 12) and Berlin Iron Bridge Company.

Deflections of beams also are treated formally by various authors but are empirically treated by Kidder. According to Kidder (1886, p. 326), for rolled-iron beams, the deflection in inches can be approximately calculated by dividing the square of the span in feet by 70 times the depth of the beam (in inches).

According to Stoney (1873, article 431), the shearing stress in the web is calculated as a simple quotient of the shear force and the web area. In article 432, he goes on to explain that the combination of normal and shear forces in the web of plate girder bridges creates uncertainty over the direction of the stress (strains). Hiroi derives a conventional form of equation for shearing stress but proposes designing girder webs to carry the entire shear as an average quantity (1893, p. 24). He also describes the design of stiffeners as the process of designing a conventional column inclined at 45 degrees by Gordon's formula (see Chapter 9), using a numerator of 8,000 lb/in.^{2} and a modifier of the height/thickness ratio of 3,000. Hiroi presents the results of a general analysis in tabular form, giving the allowable web shearing stress for various ratios of height of girder web/thickness (see Figure 8-3).

From Charles Haslett and Charles Hackley (1859, p. 211) a rectangular cast-iron bar will bear a central weight in pounds equal to the constant 1,490 times the width (inches) times the depth (inches) squared divided by the span in feet. Humber (1869) says that the best dimensions of a cast-iron beam have a bottom flange area of six times the top flange area (due to the difference between compressive and tensile capacity). Thus, the total distributed breaking weight is equal to the area of the bottom flange in inches times the depth of the beam in inches, divided by the span in feet. This can be compared to Fleeming Jenkin's (1873, p. 294) account of Eaton Hodgkinson's rule: $M = 16{,}500$ times the area of the tension flange (inches squared) times the depth (inches) for which the implied maximum stress equals

27

$\frac{h'}{b}$	c	$\frac{h'}{b}$	c
40	3870	70	1880
42	3680	75	1690
44	3500	80	1520
46	3320	85	1380
48	3150	90	1250
50	3000	95	1140
52	2860	100	1040
54	2700	110	880
56	2590	120	760
58	2470	130	650
60	2350	140	570
65	2100		

Figure 8-3. Hiroi's table of allowable girder web stress.
Source: Hiroi (1893), p. 27.

8.25 tons/in. squared (breaking, or ultimate, stress). Kidder (1886, p. 307) reiterates Hodgkinson's rule for the design of cast-iron beams for buildings that the breaking load in tons is equal to the constant 2.166 times the area of the tension flange (inches) times the depth of the beam (inches) divided by the span in feet. Based on the same assumption concerning distribution of material to the flanges, Haslett and Hackley's (1859) implied maximum (breaking) stress is then 6.4 tons/in.^{2}, indicating a more conservative approach, or allowing the use of lower grades of iron.

Most of the procedures prescribed for simply supported cast-iron beams rely on empirical formulas (see Chapter 4). In the description from Haslett and Hackley, having given (on p. 205) the precise method of calculating bending moments (strains) on a beam, the total concentrated breaking load is found from basic geometry and a generic value of maximum tensile stress (p. 212). This rule is said to apply to cast-iron beams with the bottom flange area equal to six times the top flange area. "Multiply the sectional area of the bottom flange in square inches by the depth of the beam in inches, and divide the product by the distance between the supports, measured in feet, then 2.14 times the quotient will give the breaking weight in tons." The authors also refer to previous experiments in which the tensile strength of cast iron is found to be 1/36 the compressive strength.

Floor planks, spanning across rafters or purlins, are an instance of continuous beams, although they are generally treated as simply supported for simplicity. For mill building floors, various arched and reinforced floor types are described in Ketchum (1903, pp. 249–250). These include arched floors of various forms, concrete floors with expanded metal reinforcement, and patent floors, such as the Roebling Fireproof floor system and the Buckeye Fireproof Floor. For arched floors, Ketchum recommends a span/rise ratio of eight or less. His formula for the thrust exerted by an arch floor is

$$T = 1.5\, WL^2/R$$

where

T is the calculated horizontal thrust in lb/lineal ft,
W is the floor load in lb/ft^2,
L is the span in ft, and
R is the rise in in.

Wood plank floor spans are sized based on a table of Ketchum's (1907, p. 297), reproduced as Table 8-1. It is possible to reconstruct the calculations that produced this table, using a bending moment of $wL^2/8$ and a working stress of 400 lbs/in.^{2} for spruce or white pine, and of 500 lbs/in.^{2} for yellow pine. For instance, for a 100 lbs/ft^2 load and an 8-ft span, Ketchum recommends a thickness of 3.4 in. Using these values, the actual stress for this floor is approximately 400 lbs/in.^{2} The working stress for the remaining entries in the table is similar.

Continuous Girders

A relatively common industrial use of continuous girders is a multibay crane girder, an example of which is shown in Figure 8-4. Continuous girders were often analyzed by

TABLE 8-1. *Excerpt: Plank Floor Sizes (Spruce or White Pine)*

	Thickness, in Inches, for Various Loads per Square Foot of Plank[a]										
Span in ft	lb. / 30	lb. / 40	lb. / 50	lb. / 75	lb. / 100	lb. / 150	lb. / 175	lb. / 200	lb. / 225	lb. / 250	lb. / 300
4	0.9	1.1	1.2	1.7	1.9	2.1	2.2	2.4	2.5	3.1	2.9
5	1.2	1.4	1.5	2.1	2.4	2.6	2.8	3.0	3.2	3.8	3.7
6	1.4	1.6	1.8	2.6	2.9	3.1	3.4	3.6	3.8	4.6	4.4
7	1.7	1.9	2.1	3.0	3.3	3.7	3.9	4.2	4.5	5.4	5.2
8	1.9	2.2	2.4	3.4	3.8	4.4	4.5	4.8	5.1	6.1	5.9
9	2.1	2.5	2.7	3.9	4.3	4.7	5.1	5.4	5.8		—
10	2.4	2.7	3.1	4.3	4.8	5.2	5.6	6.1			—
11	2.6	3.0	3.4	4.7	5.3	5.8					—
12	2.9	3.3	3.7	5.2							—
13	3.1	3.6	4.0	5.6							—
14	3.4	3.9	4.3	6.1							—

[a]For yellow pine use 9/10 of the thickness.
Source: Adapted from Ketchum (1907).

Figure 8-4. A continuous crane girder in the Pennsylvania Railroad Juniata Locomotive Shops (1889).
Source: Photograph by Sikander Porter-Gill.

assuming or finding a point of inflection based on a set of simplifying assumptions. This is certainly the approach taken by Francis Webb Sheilds (1871), who recommends assuming a hinge at three-fourths the length of the outer span and treating the remainder of the span as if it were a cantilever projecting from the middle span. Humber (1869, article 29, p. 12) says that continuous beams can always be considered as a combination of fixed/fixed and fixed/pinned. The main objective of his analysis seems to be to locate the inflection point. These calculations are followed by some approximations for variable loading. Humber initially calculates fixed-end moments for a "beam of uniform strength," which means a beam with a section modulus that varies as a parabola. This is a more accurate representation of many girder types (e.g., bowstring) than the simpler constant moment of inertia. According to Sheilds, the fixed/fixed uniformly loaded moments for such a beam are $wL^2/6$. Humber's further equations for moments in a multispan girder (article 33, pp. 14–15) must be considered approximate, because they use either fixed/pinned moments (at exterior piers) or fixed/fixed moments (at interior piers), varying the load placement to produce maximum negative or positive bending moment. He continues to use $wL^2/6$ as the maximum moment due to movable load while taking 2/21 wL^2 as the maximum support moment for the dead load on a multispan beam for negative moment calculations and 3/32 wL^2 for positive moment calculations. This is certainly close to the calculated value on a four-span beam, but Humber considers this correct for all of the interior supports for an arbitrary number of piers (although an exception is made for three-span girders). Humber's use of values for a beam of uniform strength where the beam undergoes moment reversals is questionable. However, his approximations of locations of points of inflection have value for bridge designers.

Jenkin (1873, p. 301) also uses the location of an inflection point and the calculation of moments in a statically determinate beam as a means of analyzing a two-span continuous girder. According to him, continuity is rarely worth the trouble, and when continuous girders are built they are of no more than two spans.

Hermann Haupt (1858, p. 102) says the following regarding continuous girder bridges, speaking primarily of wooden Howe trusses:

> When a beam is laid over several supports, its strength for a given interval is much greater than when simply supported at the ends. The same principle is applicable to bridges, and when several supports occur in succession, it is of great advantage to continue the upper and lower chords, if the bridge is straight, across the piers. By this arrangement, the strength of chords of each central span in a series would be double that of the same spans disconnected, and the extreme spans would be stronger in the proportion of 3 to 2.

The most widespread discussion of the analysis of multiple span bridges concerns the application of Clapeyron's theorem, credited to Emile Clapeyron, also called the three-moment equation, to the analysis of continuous girders. According to Merriman (1876), this is the three-moment equation in its most general form. It is applied to the arbitrary number of spans shown in the diagram in Figures 8-5 and 8-6.

$$M_{r-1}l_{r-1} + 2M_r(l_{r-1} + l_r) + M_{r+1}l_r = P_{r-1}l_{r-1}^2(k - k^3) + P_r l_r^2(2k - 3k^2 + k^3)$$

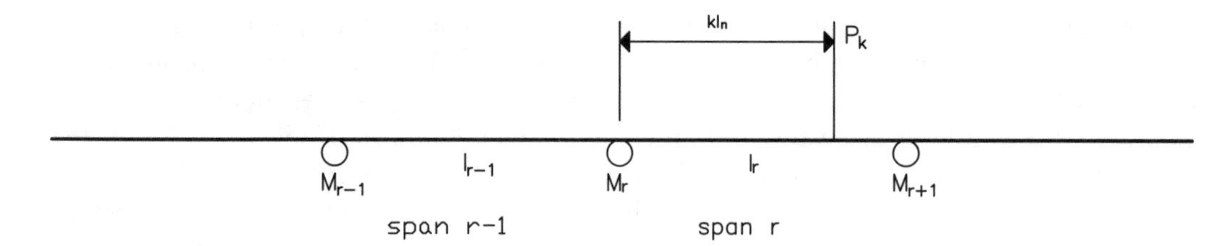

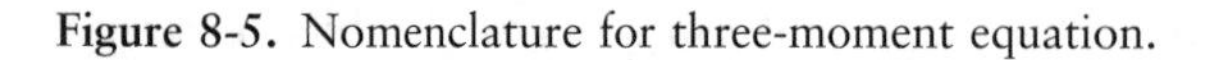
Figure 8-5. Nomenclature for three-moment equation.

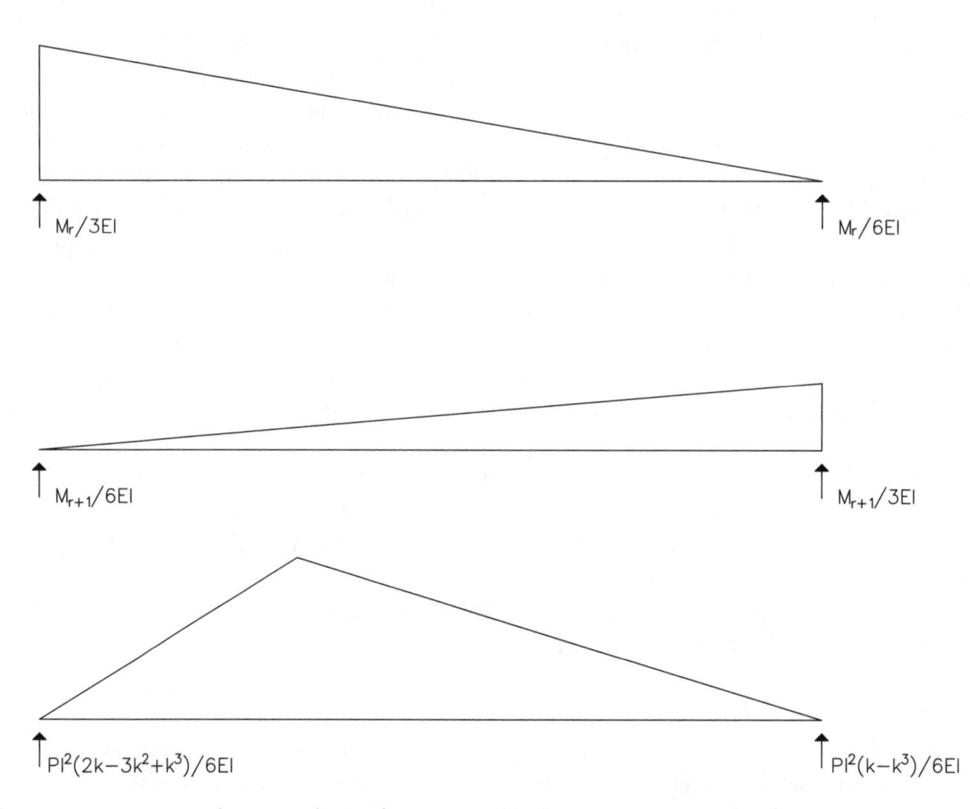

Figure 8-6. Conjugate beam analysis of a span in the three-moment equation.

This expression comes from equating the slope at support *r* by expressing it as the result of forces to the left and forces to the right of the support. Based on the diagrams in Figure 8-6, the slope components can be found.

At the left support

$$6EIt_r = -2M_r l_r - M_{r+1} l_r + P_r l_r^2 (2k - 3k^2 + k^3)$$

At the right support

$$6EIt_r = M_{r-1} l_{r-1} + 2M_r l_{r-1} - + P_{r-1} l_{r-1}^2 (k - k^3)$$

In these equations, which are formulated at a support numbered r, the moment M, the span length l, the slope t, and the load P are indexed to the support r or the span r to the right. The ratio k represents the proportion of the span (measured from the left) at which the load P_k is found. Looking, for instance, at Merriman's span r, from support r to $r + 1$, with a load P, applied at distance kl from support r, the slope at r has three components, the slope due to M_r, to M_{r+1}, and the slope due to the applied load P. Using the conjugate beam diagrams in Figure 8-6 for each of these three components results in the aforementioned equations.

This equation can be viewed as a direct application of the conjugate beam method. The coefficients in the third part of Figure 8-6 are found from a direct calculation of the reactions on the conjugate beam. The bending moment at $kl = P(k)(1 - k)l$. The moment area is thus $(1/2)Pl^2k(1 - k)$. Based on the formula for the centroid of triangle, the right reaction of the moment area is $(1/2)Pl\ k(1 - k)(1 + k)$. The slope equation for the left side can be obtained by substituting $1 - k$ for k, or by repeating this derivation for the geometry of the moment area. At least for single concentrated loads in the spans, the three-moment equation as given can be derived from the coefficients in this diagram by combining the moments for a single support point and eliminating $6EI$ between the equations.

A. Jay Du Bois (1888) reaches the same conclusions as Merriman, although his sign convention is different from Merriman's. Du Bois takes downward directed loads, i.e., A and B, to be negative, whereas Merriman makes them positive, hence, an opposite sign convention. Merriman (and Du Bois) supplement the three-moment equation with a solution method that allows the direct determination of the moments at the supports for any single span loaded condition, using beam coefficients, constant for any beam configuration, that are computed from span and stiffness, and a pair of coefficients for the loading condition at the two ends of the loaded span. For dead load plus live load cases, the results of such an analysis must be superposed to obtain the full bending moments. This requires recourse to tabular computations of the support moments in a continuous girder. An application of this procedure is described in Box 8-1.

Kidder, in speaking about the strength and stiffness of continuous girders in building construction (1886, pp. 327–335), considers only the two-span and three-span cases and practically considers only uniformly distributed loading. He makes no use of the three-moment equation or any of the formulas based on support moments developed and used by the bridge engineers of the time but begins with the calculation (by formula) of the support reactions and proceeds to the calculation of the support moments. Kidder concludes that a two-span beam is no stronger than a simply supported beam but that a three-span beam offers a 25% increase in strength. He does highlight the considerable gain in stiffness by the use of a continuous beam. His methodology in the handbook is the simple presentation of formulas, however; he has derived the formulas in a separate article.

Similar to other types of structure previously discussed, the methods used for the analysis of continuous beams varied from empirical formulas through devices to make the computation manageable, such as the assumption of hinge locations, to complex and detailed calculations of the forces in the girder. In most cases, the standard approach to continuous girder bridges appears to have been to use them only where necessary, for instance, in movable span structures. In building structures, approaches to analysis and design appear to be mostly the application of approximate methods or simplifying formulas.

Box 8-1

The following example is a modification of Du Bois's (1888) Example 3. A continuous bridge truss has a constant moment of inertia and five spans of 60 ft, 80 ft, 80 ft, 80 ft, and 60 ft. The supports are numbered 1 through 6. Dead load is 800 lbs/ft in all spans and live load of 1,200 lbs/ft in span 2 and 3. The method of undetermined coefficients, and its application to solutions of the three-moment equation, is thoroughly described in Merriman (1876). The pattern for the coefficients c (representing loaded span effects on spans to the left of the load) and d (representing spans to the right of the load) is given, up to an index of 5. The calculation of the coefficients for each of the spans proceeds thus:

$c_1 = 0$

$c_2 = 1$

$c_3 = -2(l_1 + l_2)/l_2 = -3.5$

$c_4 = -2c_3(l_2 + l_3)/l_3 - c_2(l_2/l_3) = 13$

$c_5 = -2c_4(l_3 + l_4)/l_4 - c_3(l_3/l_4) = -48.5$

$d_1 = 0$

$d_2 = 1$

$d_3 = -2(l_5 + l_4)/l_4 = -3.5$

$d_4 = -2d_3(l_4 + l_3)/l_3 - d_2(l_4/l_3) = 13$

$d_5 = -2d_4(l_3 + l_2)/l_2 - d_3(l_3/l_2) = -48.5$

($d_i = c_i$ for symmetric span configurations)

In a bridge consisting of s spans, numbered 1 through s, and $s + 1$ supports, numbered 1 through $s + 1$, when span r is the loaded span, it is found (Merriman 1876) that the moment at support n, for $n < r + 1$, is

$$M_n = c_n \frac{Ad_{s-r+2} + Bd_{s-r+1}}{d_{s-1}l_2 + 2d_s(l_1 + l_2)}$$

and for $n > r$

$$M_n = d_{s-n+2} \frac{Ac_r + Bc_{r+1}}{c_{s-1}l_{s-1} + 2d_s(l_{s-1} + l_s)}$$

The coefficients c_i and d_i are determined as shown. A and B are loading coefficients, both equal to $wl^3/4$ for uniformly distributed loads and equal to the coefficients in P defined in Figure 8-6 for concentrated loads. In this case, for $r = 1$ (span 1 loaded), $A = B = 1/4\, wL_1^3 = 27/256\, wL_2^3$, M_5 is calculated first, and the remaining moments are found easily, being proportional to M_5.

$$M_2 = d_5 M_5 = 2{,}619/93{,}568\, wL_2^2$$

$$M_3 = d_4 M_5 = -702/93{,}568\, wL_2^2$$

$$M_4 = d_3 M_5 = +189/93{,}568\, wL_2^2$$

$$M_5 = (27/256\, wL_2^3)/[13\, L_2 + 2\,(-48.5)1.75\, L_2] = -(27/256)(4/731)\, L_2 = -54/93{,}568\, wL_2^2$$

$r = 2$ (span 2 loaded) $A = B = 1/4\, wL_2^3$

M_2, being to the left of the loaded span, is calculated differently in this case.

$$M_2 = -(71/8\ wL_2^3)/[13\ L_2 + 2\ (-48.5)1.75\ L_2]$$
$$= (71/8)(8/1,254)\ L_2 = 142/2,508\ wL_2^2$$

The remaining support moments are calculated as for $r = 1$.

$$M_3 = d_4 M_5 = 130/2,508\ wL_2^2$$

$$M_4 = d_3 M_5 = -35/2,508\ wL_2^2$$

$$M_5 = -(5/8\ wL_2^3)/[13\ L_2 + 2\ (-48.5)(1.75\ L_2)]$$
$$= -(5/8)(8/1,254)\ L_2 = 10/2,508\ wL_2^2$$

$r = 3$ (span 3 loaded) $A = B = 1/4\ wL_2^3$

$$M_2 = (19/8\ wL_2^3)/[13\ L_2 + 2(-48.5)(1.75\ L_2)]$$
$$= (19/8)/(-627/4) = -38/2,508\ wL_2^2$$

$$M_3 = 133/2,508\ wL_2^2$$

It is unnecessary to calculate M_5 in this case, as it is equal to M_2 by symmetry.

Table B8-1-1 presents loaded spans and support moments results. Following Merriman (1876), positive support moments result adjacent to the loaded span, opposite to the conventional modern sign convention for bending moments. Table B8-1-2 shows the dead load moments that result at each support for a dead load of 800 lbs/ft. The moment is expressed in lbs/ft. Any case of a fully loaded span can be solved using these coefficients, as in the example in Table B8-1-3. The first column is the summation of the rows of the previous table, showing the dead load moments resulting from all spans loaded. In this case, the live load has been placed in span 2 and span 3. The final result in the last column is obtained by multiplying the previous column by L_2^2 (6,400 ft^2).

TABLE B8-1-1. *Support Moments for Various Span Loading Conditions*

	M/wL_2^2				
Support	Span 1 Load	Span 2 Load	Span 3 Load	Span 4 Load	Span 5 Load
2	2,619/93,568	142/2,508	−38/2,508	10/2,508	−54/93,568
3	−702/93,568	130/2,508	133/2,508	−35/2,508	189/93,568
4	189/93,568	−35/2508	133/2,508	130/2,508	−702/93,568
5	−54/93,568	10/2,508	−38/2,508	142/2,508	2,619/93,568

TABLE B8-1-2. *Dead Load Moments: Five-Span Bridge Example*

	Dead Loads				
Support	$D.L.M/L_2^2$ span 1 (lbs/ft)	$D.L.M/L_2^2$ span 2 (lbs/ft)	$D.L.M/L_2^2$ span 3 (lbs/ft)	$D.L.M/L_2^2$ span 4 (lbs/ft)	$D.L.M/L_2^2$ span 5 (lbs/ft)
2	22.3	45.3	−12.1	3.2	−5.0
3	−5.9	41.4	42.4	−10.8	1.6
4	1.6	−10.8	42.4	41.4	−5.9
5	−0.5	3.2	−12.1	45.3	22.3

TABLE B8-1-3. *Dead Load and Live Load Moments: Five-Span Bridge with Spans 2 and 3 Loaded*

	Load Case—Spans 2 and 3 Loaded: 1,200 lbs/ft							
Support	$TDLM/L_2^2$ (lbs/ft)	$L.L.M/L_2^2$ Span 1 (lbs/ft)	$L.L.M/L_2^2$ Span 2 (lbs/ft)	$L.L.M/L_2^2$ Span 3 (lbs/ft)	$L.L.M/L_2^2$ Span 4	$L.L.M/L_2^2$ Span 5	Total Moment (lbs/ft)	Total Moment ton ft
2	58.2		67.9	18.0			144.1	461
3	68.7		62.2	63.7			194.6	623
4	68.7		–16.7	63.7			115.7	370
5	58.2		negligible	18.0			76.2	244

While the application of analytical methods is necessary to the development of continuous girders, simply supported girders can be designed by empirical or semi-empirical methods. The use of such methods persisted well into the late nineteenth century, but the design of practically all girders was by analytical methods by the end of the century.

References Cited

Bender, C. (1876). Application of the theory of continuous girders to economy in bridge building, *J ASCE*, 1(5), 142–198.

Bow, R. H. (1874). *A treatise on bracing*. Van Nostrand, New York (originally published in Britain in 1851).

Campin, F. (1868). *On the construction of iron roofs*. Van Nostrand, New York.

Du Bois, A. J. (1888). *The strains in framed structures*, 4th Ed. John Wiley and Sons, New York.

Haslett, C., and Hackley, C. (1859). *Mechanic's, machinist's, and engineer's practical book of reference*. W.A. Townsend, New York.

Haupt, H. (1858). *General theory of bridge construction*. D. Appleton, New York.

Hiroi, I. (1893). *Plate-girder construction*, 2nd Ed. Van Nostrand, New York.

Humber, W. (1869). *A handy book for the calculation of strains*. Van Nostrand, New York.

Jenkin, F. (1873). *Bridges: an elementary treatise on their construction and history*. Adam and Charles Black, Edinburgh.

Ketchum, M. (1903). *Design of steel mill buildings*. Engineering News Publishing, New York.

Ketchum, M. (1907). *Design of steel mill buildings*, 2nd Ed. Engineering News Publishing, New York.

Kidder, F. (1886). *The architects' and builders' pocket-book*, 3rd Ed. John Wiley and Sons, New York.

Merriman, M. (1876). *On the theory and calculation of continuous girders*. Van Nostrand, New York.

Nicholson, P. (1826). *Practical carpentry, joinery, and cabinet-making*. Thomas Kelly, London.

Sheilds, F. W. (1871). *Treatise on iron construction*, 2nd Ed. John Weale, London.

Stoney, B. B. (1873). *The theory of strains in girders and similar structures*. Van Nostrand, New York.

Swain, G. F. (1896). *Notes on the theory of structures*, 2nd Ed. Mimeographed lecture notes. Massachusetts Institute of Technology, Department of Civil Engineering, Cambridge, MA.

Tredgold, T. (1820). *Elementary principles of carpentry*. J. Taylor, London.

Waddell, J. A. L. (1894). *The designing of ordinary iron highway bridges*, 5th Ed. John Wiley and Sons, New York.

Yeomans, D. T. (1987). "Designing the beam: from rules of thumb to calculations." *J Inst Wood Sci*, 11(61), 43–49.

9

Analysis of Columns

The procedures generally used in the analysis of columns in the late 1800s were developed in part by Eaton Hodgkinson (1846) on the basis of his experiments on iron columns. Essentially empirical, this work fitted column strength data, obtained from tests on small iron specimens, to a logarithmic curve. Different curves were determined for differing support conditions and for differing materials. Lewis D. B. Gordon, using Hodgkinson's published data, arrived at empirical equations, which William John Macquorn Rankine (1877) verified by theoretically determining the point at which an eccentrically loaded column reaches breaking stress due to a combination of axial force and bending. Because of its basic simplicity for calculations, this formulation became widely used in engineering practice in the United States and was known as Gordon's formula. Authors such as William Merrill (1870) identify the Rankine-Gordon formula as being a simplification of Hodgkinson's curves: although this may be partially true of Gordon's formula, Rankine's development of the same equation proceeds from an entirely theoretical basis. Details of these formulas are given in the following.

Iron Columns

Cast-iron, and later wrought-iron columns were widely used in bridge and commercial and industrial building in the United States. Figure 9-1 shows an example of a cast-iron column in an 1882 industrial building for the Phoenix Iron Company, which also produced sectional wrought-iron columns for use in bridges. Figure 9-2 shows a typical

Figure 9-1. Cast-iron column at Phoenix Iron Company foundry building, 1882 (HAER PA,15-PHOEN,4A–18).
Source: Photograph by Jet Lowe.

metal (wrought iron or steel) building column, which is assembled by stitching plates and angles, or in this case channels and lattice bars, together with rivets applied in the shop. While Hodgkinson's work is recognized in the development of column formulas, other American authors circulated this information in various forms. Merrill (1870), for instance, presents a thorough summary of Hodgkinson's findings and applies both Gordon's and Hodgkinson's formulas to the interpretation of the results. He presents a pair of three-dimensional plots that integrate the results of Hodgkinson's small-scale tests into a set of results for use with cast-iron columns of various configurations, solid cylindrical and hollow cylindrical with square ends or rounded ends. Merrill is principally concerned with the use of wrought iron for ties, or tension members, and the use of cylindrical struts of cast iron, either solid or hollow. As such, his main interest in column theory is the work of previous writers on cylindrical sections. He reviews the results of Hodgkinson's series of tests on solid cast-iron bars, constructed with either square or rounded ends. He notes the difference between very short bars, which have a breaking weight close to the compressive strength of the material, and longer bars, whose breaking weight is governed by buckling. An example presentation of Hodgkinson's experiments is shown in Figure 9-3. While the results for solid bars are taken more or less directly from the results of Hodgkinson's testing, the breaking

Figure 9-2. Lattice wrought-iron or steel columns at Juniata Locomotive Shops, Altoona, PA, 1889.
Source: Photograph by Sikander Porter-Gill.

weight of a hollow cylindrical bar is calculated as the difference between the breaking weight of a solid bar with a diameter equal to the outside diameter and a bar with diameter equal to the inside diameter. Merrill presents Hodgkinson's and Gordon's formulas for a solid cylindrical cast-iron bar. Hodgkinson's formula is in two parts, a formula for the strength of a long bar and a correction for shorter bars. For a long bar, with rounded ends, Merrill presents Hodgkinson's formula as

$$W = 33,380\frac{d^{3.76}}{l^{1.7}}$$

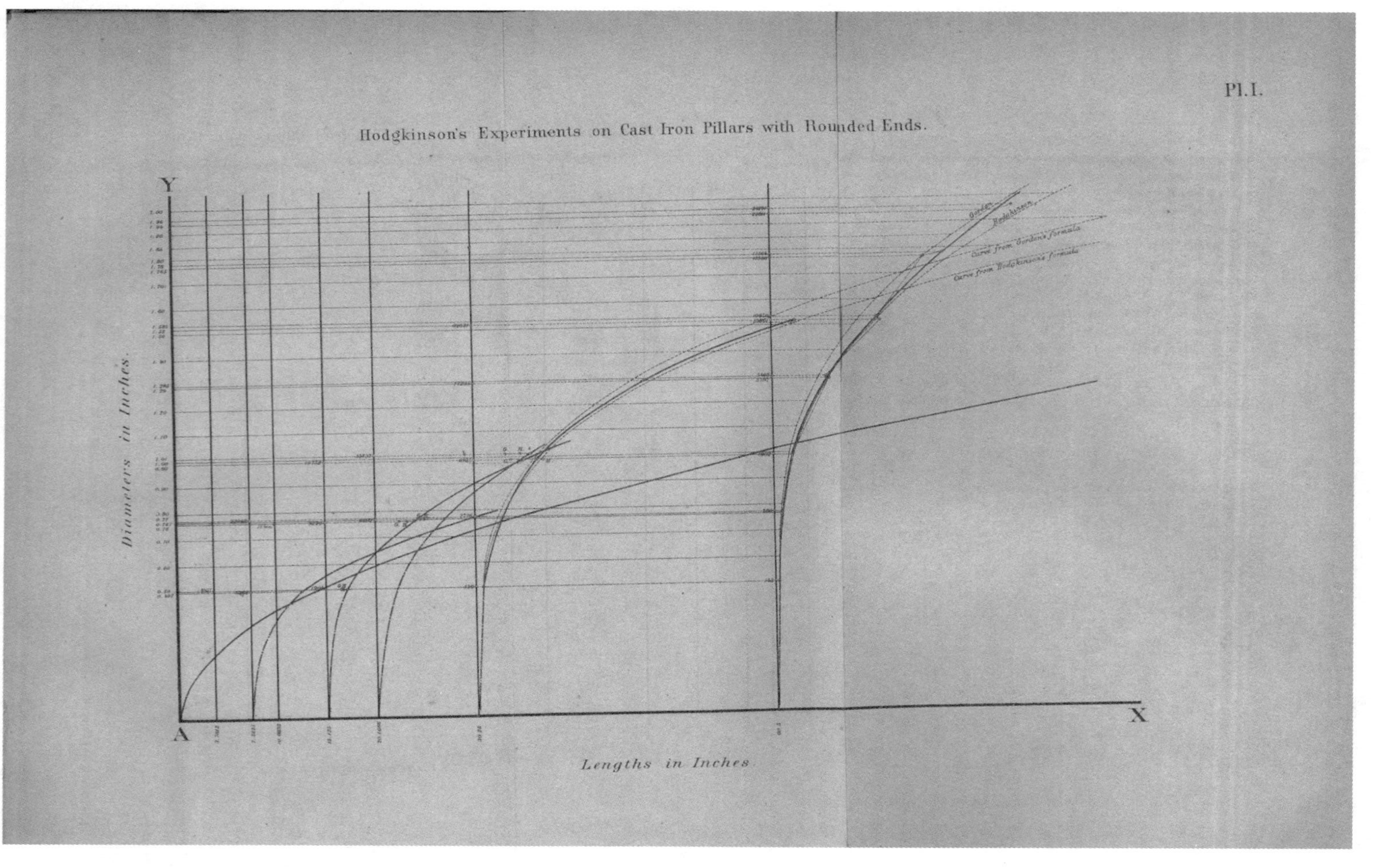

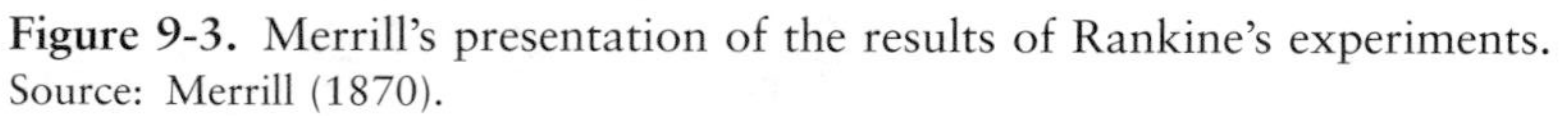

Figure 9-3. Merrill's presentation of the results of Rankine's experiments.
Source: Merrill (1870).

The coefficient is modified to 98,900 for flat ends, approximately tripling the breaking load. In this version of Hodgkinson's formula, W is the breaking weight of the section, d is the diameter in inches, and l is the length of the column or strut in feet.

Plotting breaking weight versus l/d using this empirical formula results in a monotonically decreasing hyperbolic curve, where the breaking weight of the column decreases with increasing l/d ratio. Factoring out the dependence of the stress on the square of the diameter transforms this equation into an expression for buckling stress, found to be inversely proportional to $(l/d)^{1.7}$. For shorter columns, the inelastic buckling effects require a correction formula for l/d less than 15 for columns with rounded ends and less than 30 for columns with square ends. The actual breaking weight is given as

$$\frac{bc}{b+\frac{3}{4}c}$$

where b is the breaking weight calculated by the aforementioned buckling formula and c is the crushing strength of the material in pounds per square in. based on a crushing stress of 100,000 to 120,000 lbs/in.^{2}

A crushing strength of 100,000 lbs/in.^{2} is used in the calculations shown in Table 9-1, which compares the results of the application of Hodgkinson's and Gordon's formulas. The application of Hodgkinson's formula requires the development of a table with nine columns of calculations and the use of a table of logarithms to evaluate separately the exponents of the length and the diameter. The further calculation of a correction factor for shorter struts is also required, such as a 5-ft strut 4 in. in diameter ($l/d = 15$). For a hollow section, this calculation would have to be completed twice.

Gordon's formula takes the form of a reduction in the crushing strength of the material based on the characteristics of end conditions, slenderness ratio, material, and bending stiffness of the cross section. In the most general form, as presented by A. Jay Du Bois (1887, p. 355), the formula is written as follows:

$$P = \frac{\mu}{1+c\frac{l^2}{d^2}}$$

TABLE 9-1. *Comparison of Tabular Calculation of Column Strength by Hodgkinson's and Gordon's Formulas*

Hodgkinson Column Calculations (Nine-column calculation required)

Length	Log Length	Log Length ×1.73	Diameter	Log Diameter	Log Diameter ×3.76	Column breaking weight	Correction	Column breaking weight
10	1	1.73	6	.778	2.92	1,627,000	—	1,627,000

Gordon Column Calculations

Length	Diameter	$(l/d)^2$	Denominator	Column breaking weight
10	6	400	1.267	2,232,000

In this equation, μ is the breaking strength of the material, *l*/*d* is the height/diameter ratio of the column, and *c* is an empirically determined constant. Typical values for iron struts with pinned ends are μ = 40,000–80,000 (lbs/in.^{2}), *c* = 1/2,250 for wrought iron, 1/400 for cast iron, rounded (pinned) ends, and the value of *c* is doubled for flat (fixed) ends. It is noted that the radius of gyration *r* is approximately equal to *d*/2 for hollow, cylindrical shapes. The figure in Du Bois (1888, p. 344), which is reproduced in Figure 9-4, summarizes these values for wrought-iron shapes.

According to Merrill (1870), "Gordon's formula appears to be an adaptation for engineers who are not familiar with the use of logarithms." (p. 28) The formula certainly had simplicity in its favor. The calculation of column breaking weight according to Gordon's formula requires no more than a five-column calculation consisting of addition and multiplication (see Table 8-1): the only lookup might be in using a Crelle (1897) table for multiplication. The formula also captures the double curvature of the curve relating slenderness

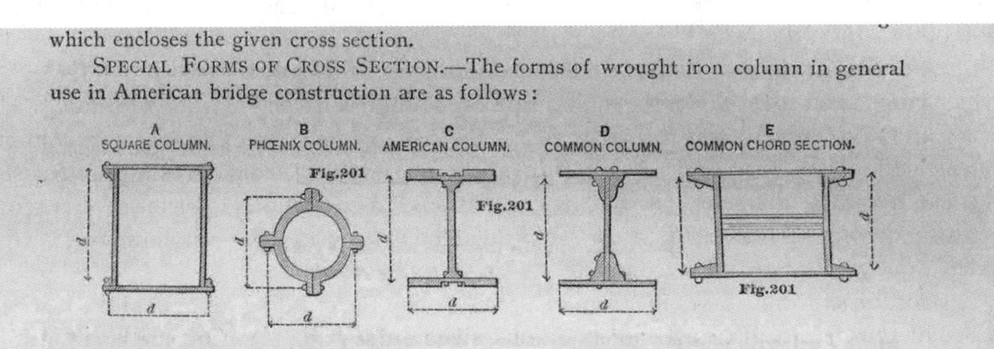
which encloses the given cross section.

SPECIAL FORMS OF CROSS SECTION.—The forms of wrought iron column in general use in American bridge construction are as follows:

For these forms, the following special formulæ are recommended by C. Shaler Smith for *wrought iron;* where *d* is the least dimension of the rectangle enclosing the cross section, and *l* is the length, both in inches.

	A.	*B.*	*C.*	*D.*	*E.*
Flat ends,	$\frac{38500}{1+\frac{l^2}{5820\,d^2}}$,	$\frac{42500}{1+\frac{l^2}{4500\,d^2}}$,	$\frac{36500}{1+\frac{l^2}{3750\,d^2}}$,	$\frac{36500}{1+\frac{l^2}{2700\,d^2}}$,	The pin being so placed that the moment of inertia is, as near as practicable, equal on both sides of same, use formula for square column.
One pin end,	$\frac{38500}{1+\frac{l^2}{3000\,d^2}}$,	$\frac{40000}{1+\frac{l^2}{2250\,d^2}}$,	$\frac{36500}{1+\frac{l^2}{2250\,d^2}}$,	$\frac{36500}{1+\frac{l^2}{1500\,d^2}}$,	
Two pin end,	$\frac{37500}{1+\frac{l^2}{1900\,d^2}}$,	$\frac{36600}{1+\frac{l^2}{1500\,d^2}}$,	$\frac{36500}{1+\frac{l^2}{1750\,d^2}}$,	$\frac{36500}{1+\frac{l^2}{1200\,d^2}}$,	

The safe working stress is found by dividing the "crippling stress," as determined by the above formula, by $4+\frac{l}{20\,d}$, where *l* is length in inches, and *d* is least dimension of enclosing rectangle.

Figure 9-4. Gordon's formula adapted to various column sections.
Source: Du Bois (1888).

ratio to maximum column stress. By changing coefficients, Gordon's method can be adapted to any material—wood, wrought iron, steel—and any shape. Although often developed and presented strictly as an empirical relationship for columns, an equation equivalent in form, Rankine's formula, can be derived from theoretical considerations of column buckling. In the following section, the rational basis of Rankine's formula is explored.

The Rankine-Gordon Formula for Column Capacity

Du Bois's application of Gordon's formula for column loading is typical of the treatments of the period for iron columns of any type. Empirical constants for use in Gordon's formula are given according to the shape of the column and other characteristics of the structure. (see Figure 9-4). The ultimate strength of the material is left in the numerator or modified by dividing by a factor of safety. As such, the formula can be used to calculate an ultimate or an allowable stress for the column. Examples of such constants are shown as follows, but other constants may be used. In the material from Du Bois reproduced in Figure 9-4, the slenderness ratio is modified to l/d. Du Bois's recommended factors of safety, applied separately, range from four and up depending on the live load/dead load ratio ($4 + l/20d$ for wrought iron; in this formula l = live load and d = dead load).

The *Album of Designs*, published by the Phoenix Bridge Company (1888), gives two sets of formulas for column strength: one for ultimate strength and one for allowable column load. For ultimate strength, the company recommends $\mu = 42{,}000$ lbs/in.^{2} and $c = 1/50{,}000$ for flat-ended columns and $c = 1/30{,}000$ for rounded ends. The formulas for the Phoenix Bridge Company use l/r in place of l/d. Because, for a Phoenix column, Du Bois takes d as the distance between flange rivets, the difference between $d/2$ and r is somewhat greater than for an ordinary cylindrical column. After calculation of the maximum load on a column, a safety factor of 5 on column loads is recommended in the *Album of Designs*.

The Rankine-Gordon equation can be derived from basic principles, or it can be determined empirically. Both approaches are evident in the use of this formula in the nineteenth century. However, nearly all column data are fitted to a curve of this form. The derivation of this curve is given in John Davenport Crehore (1886, pp. 289–295) and is presented here in Box 9-1. Although the assumptions are similar to those used in the usual development of the Euler buckling load, Crehore uses the crushing strength of the material of the column as a limiting value and works with P explicitly as the limiting column load. The resulting equation, unlike a modern buckling equation, accounts for combined bending and axial force and yields the column capacity for an initially eccentrically loaded column.

In addition to Crehore's analytical treatment of the rules for column design, he derives formulas of a similar character in empirical form based on expected proportions, leaving the proportionality constants to be determined experimentally. Starting, for instance, with the proportionality of bending stress and deflection (the right-hand side being an expression for deflection)

$$B_1 \sim \frac{Pl^2}{Sh^2}$$

Box 9-1

For a pinned/pinned column, the moment at a point x distant from the lower support is

$$M_x = -EI\frac{d^2y}{dx^2} = Py$$

where P is the axial force (causing failure) and y is the lateral deflection of the column.

Calling Q the longitudinal pressure P/S and C the breaking compressive stress, the maximum moment of the internal forces must be diminished by the ratio of axial compressive stress to total stress $(C\text{-}Q)/C$. Setting the simplifying term ε^2

$$\varepsilon^2 = \frac{EI(C-Q)}{PC} = \frac{Er^2(C-Q)}{QC}$$

where r is the radius of gyration of the column. Substituting this expression into the equation for bending moments as a function of x gives

$$\varepsilon^2\frac{d^2y}{dx^2} = -y$$

This expression can be integrated with boundary conditions $y(0) = y(l) = 0$ to find

$$l = \varepsilon\left(sin^{-1}\frac{y}{a}\right)_0^0 = \varepsilon n\pi$$

Taking the least integral value $n = 1$, it is found that

$$l^2 = \pi^2\varepsilon^2 = \frac{\pi^2 Er^2(C-Q)}{QC}$$

A more modern procedure of solving the differential equation $y'' + (1/\varepsilon^2)y = 0$ and apply the boundary conditions $y(0) = y(l) = 0$ gives the same result.

Hence, the breaking force, Q, can be expressed as

$$Q = \frac{C}{\left(1+\frac{Cl^2}{\pi^2 Er^2}\right)}$$

Theoretical modifications can be made for differing end conditions, for instance, substituting $\frac{1}{4}(l/r)^2$ for $(l/r)^2$ for fixed ends.

where

B_1 is the bending stress,
P is the column load,
S is the cross-sectional area of the column,
l is the column length, and
h is the least lateral dimension of the column.

Because the total stress on the column $f = Q + B_1$, the breaking stress $Q = P/S$ can be written in terms of the total stress f, taking account of the constant of proportionality a. Thus,

$$f = \frac{P}{S} + \frac{Pl^2}{aSh^2}$$

or, factoring out P/S

$$\frac{P}{S} = \frac{f}{\left(1 + \frac{l^2}{ah^2}\right)}$$

In an equation of this form, the values of f and a are determined experimentally. The similarity to the rationally derived formula in Box 9-1 can be seen. Although this formula, according to Crehore, was first presented by Thomas Tredgold, it is known and widely used as Gordon's formula.

Regarding the effect of end-bearing conditions, Hodgkinson (1846) is more cautious than Gordon, generally presenting long column results with the square-end columns having approximately three times the capacity of a similar round-ended column, whereas applications of Gordon's formula have a coefficient of exactly 4, as dictated by buckling theory for columns. Of course, a flat-ended column is not rigidly fixed, although it can develop significant stabilizing moments at the ends, and a round-ended column is not fully pinned, as during buckling a small moment may be developed at the end of a round-ended column, so the use of a constant less than 4 is justifiable.

Bindon Blood Stoney (1873, pp. 285–286) provides additional guidance on the design of latticed columns. He simplifies a latticed column ("braced pillar") to a column subjected to the required axial force, enhanced by the application of a transverse force P at each joint between bracing and chord. This transverse force P is found to be equal to WL/R, where W is the axial force in the column, L is the column length, and R is the radius of curvature of the column due to lateral deflection. Stoney then calculates the increase in compressive stress on the inside flange and the decrease in compression (or tension) on the outside flange. He is content to show that the change in stress in the flanges and in the bracing is insignificant. He also collects data on strength of wood columns as shown in Figure 9-5.

The Gordon formula gives a theoretically consistent procedure for the determination of the buckling load of a column, which is less evident in the rules for column design in current practice. According to the steel and wood column formulas in use at the present time (American Wood Council 2006, American Institute of Steel Construction 2010), a column capacity curve has an elastic buckling part at high L/d ratios, a maximum compressive capacity at low L/d ratios, and an empirical interpolation between these two curves. Gordon's formula, whether the coefficients used are rationally or empirically based, furnishes a single column curve of reasonable accuracy for a column with given geometric shape made of a given material. As an example, in Figure 9-6 a column curve for steel is shown along with Gordon's formula specialized to an I beam, rounded ends, and mild steel, according to the theoretical ($a = 50$ kips/in.^{2}, $b = 476$) and empirical ($a = 50$ kips/in.^{2}, $b = 1200$, $(r/d)^2 = 1/10$). Gordon's formula is successful in capturing the shape of the column curve using a single formula, especially if one is willing to use empirical values of the coefficients a and b.

TABLE XII.

SOLID SQUARE PILLARS OF PINE.

DATA FROM BINDON B. STONEY'S "THEORY OF STRAINS IN GIRDERS AND SIMILAR STRUCTURES."

$h^2 = 12r^2$. Take $E = 1460000$, $C = f = 5000$.

No.	$l \div h$ Ratio of Length to Least Diameter.	Q = Breaking-Weight, in Lbs., per Square Inch.						
		Rondelet's Proportionals. Flat Ends.	Brereton's Tests. Ends in Ordinary Manner.	Gordon Formula, $Q = \frac{5000}{1 + \frac{l^2}{250h^2}}$	Hodgkinson's Formula, $Q = \frac{500Cl^2}{h^2}$	Equation (381). No End Moment.	Equation (391). One End fully fixed.	Equation (385). Both Ends fully fixed.
184	1	5000	-	-	5000	-	-	-
185	12	4167	-	3176	5000	3126	3959	4349
186	24	2500	-	1513	4940	1471	2437	3126
187	36	1667	-	809	1929	782	1485	2135
188	48	833	-	489	1085	462	960	1471
189	60	417	-	325	693	313	660	1076
190	72	209	-	230	483	221	478	642
191	10	-	1867	3571	5000	3530	4228	4529
192	20	-	1789	1923	5000	1876	2889	3530
193	30	-	1400	1087	2777	1053	1891	2581
194	40	-	1244	676	1563	653	1273	1875

Figure 9-5. Table XII reproduced from John Crehore.
Source: Crehore (1886).

Wood Columns

Wood columns also were used widely when sufficient material was available or economics dictated the use of wood. Figure 9-7 shows the use of a wood column in an 1891 Wisconsin paper mill. Referring to the table of representative wood column strengths presented in Chapter 3 (Table 3-1), we can note the divergence of the results between, for instance, Frank Kidder (1886) and Robert Maitland Brereton (1870). The principal reason for these differences is the assumption of flat ends or rounded ends, i.e., pinned conditions or fixed conditions. Having flat ends on a column has been found to increase the column strength by a factor of 4 theoretically and by a factor of 3 experimentally. The column curve of Brereton shows the column close to full strength for short *l/h* ratios and tailing off to lower values relatively slowly due to the partial fixity at the ends of the column. In contrast, a theoretical curve for a pinned end column begins at the same value for low height/thickness ratios and drops off more quickly. As the numerator in Gordon's formula is generally the breaking strength of the material in compression, this value (coefficient μ, according to Du Bois

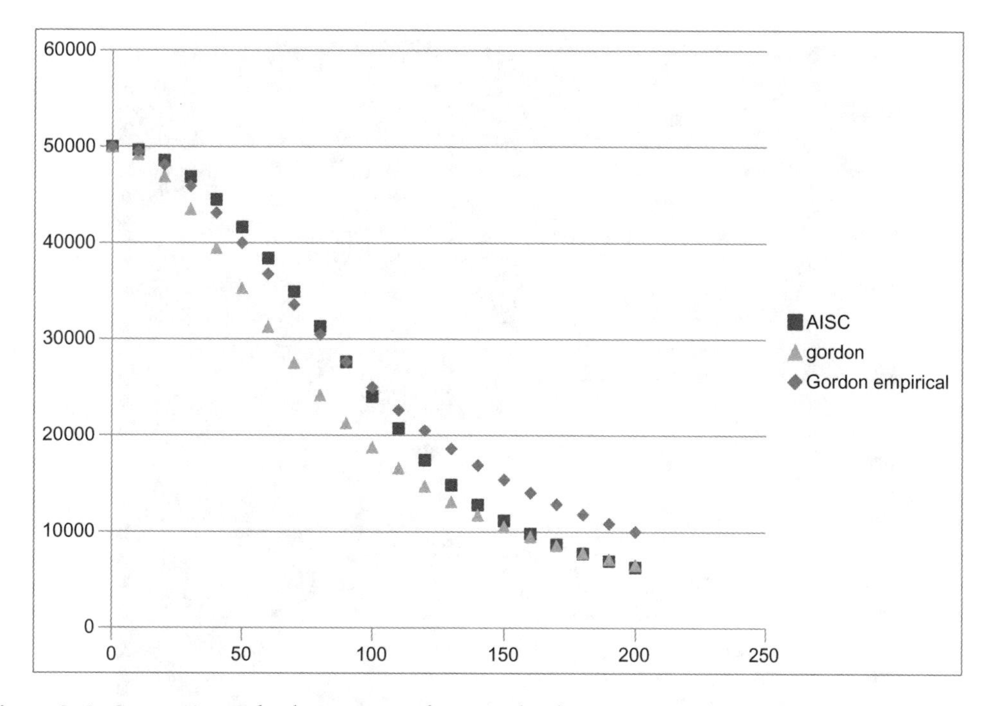

Figure 9-6. Comparison of column curves for a steel column.

[1888]; coefficient *f*, according to Crehore [1886]) is unaffected by the end conditions; the other coefficient ($1/c$ or *a*) varies by a factor of 2 to 4. According to Crehore (Table XII, presented here as Figure 9-5), Brereton's results or Rondelet's rules correspond to a value of *f* of 5,000 psi and to a value of 250 of *a* (using the l/h ratio). It is of interest to investigate the theoretical coefficients presented by Crehore (see Box 9-1). The application of this theory leads to a value of $f = 1{,}700$ (120 tons of 2,000 lbs divided by 144 in.^{2}) and a value of *a* of 2,900 (as the coefficient for $(l/r)^2$, or 242 for the coefficient of $(l/d)^2$ ($E = 1{,}500{,}000$ lbs/in.^{2}). Curves for these values are plotted in Figure 9-8. Although considering the constants in Gordon's formula as empirical parameters yields a slightly better fit to Brereton's reported results, the application of the theoretical formula of the same form, using plausible values of the modulus of elasticity and a reasonable range of 3 to 4 for the increase in *b* due to square ends, results in a reasonable fit.

The Chicago building code (City of Chicago 1905) covers four different species of timber and presents points on column curves for these materials: white pine and spruce, loblolly yellow pine, and white oak (Table 9-2). The values from the Chicago building code for square ends display the usual caution of a building code, decreasing the load for long columns more than either the theory or experiment justify. However, the design values for long columns are still greater than the values that would be used in contemporary design, which take no account of the increase in column strength afforded by square ends. The values for white pine and spruce are given in Chapter 3. The values for the other two materials are presented here. Although in form design values greatly resemble Rondelet's rules, presented in Chapter 3, Rondelet's rules are much more conservative, reducing overall

Figure 9-7. Whiting-Plover paper mill, Whiting, WI, Building No. 1., 1891 (HAER WIS,49-WHIT,1C–2).
Source: unknown.

TABLE 9-2. *Chicago Building Code (City of Chicago 1905) Design Rules for Squared-End Wood Posts*

L/d	Reduction in full strength	White Pine/Spruce	Loblolly Yellow Pine
0–15	1	1,000	750
15–30	7/8	875	650
30–40	3/4	750	560
40–45	5/8	625	460
45–50	1/2	500	375

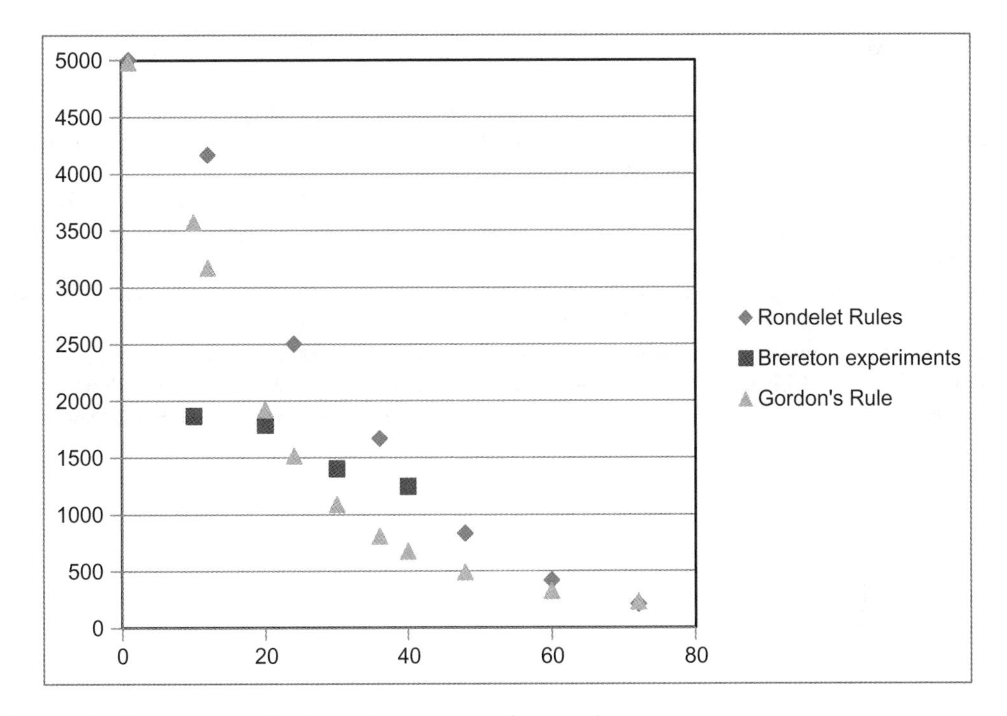

Figure 9-8. Comparison of column design curves for wood.

strength to half at an *L/d* of 24, compared with 45 for the Chicago building code. Apparently, the Chicago building code represents an application of Rondelet's rules to a square-ended column, supposing an effective length factor of approximately 2 for a flat-ended timber column.

For steel and wood column design, the analytical procedures used in the late nineteenth century were based on experimental results. The procedures used were appropriate combinations of the application of empirical rules and the application of analytical techniques to the design of columns. The widespread use of the Rankine-Gordon formula is particularly noteworthy as it resulted, especially for iron structures, in a theoretically defensible and consistent means of designing columns of all materials, lengths, configurations, and support conditions.

References Cited

American Institute of Steel Construction. (2010). *Specification for structural steel buildings*. AISC, Chicago.

American Wood Council. (2006). *National design specification for wood structures*. American Forest and Paper Association, Washington, DC.

Brereton, R. P. (1870). Discussion of Gaudard, on the present state of knowledge as to the strength and resistance of materials. *Minutes of the Proceedings of the Institution of Civil Engineers*, 1(29), 25–97.

City of Chicago. (1905). *An ordinance relating to the department of building*. Moorman and Geller, Chicago.

Crehore, J. D. (1886). *Mechanics of the girder*. John Wiley and Sons, New York.

Crelle, A. L. (1897). *Dr. A.L. Crelle's calculating tables and their application to the multiplication and division of all numbers above one thousand*. David Nutt, London.

Du Bois, A. J. (1888). *The strains in framed structures*, 4th Ed. John Wiley and Sons, New York.

Hodgkinson, E. (1846). *Experimental researches on the strength and other properties of cast iron*. John Weale, London.

Kidder, F. (1886). *The architects' and builders' pocket-book*, 3rd Ed. John Wiley and Sons, New York.

Merrill, W. (1870). *Iron truss bridges for railroads*. Van Nostrand, New York.

Phoenix Bridge Company. (1888). *Album of designs*. J. B. Lipincott, Philadelphia.

Rankine, W. J. M. (1877). *A manual of applied mechanics*, 9th Ed. Charles Griffin, London.

Stoney, B. B. (1873). *The theory of strains in girders and similar structures*. Van Nostrand, New York.

10

Analysis of Portal Frames

A portal frame is a type of rigid frame used to resist the collected wind loads on a structure. The most common use of this system is to solidify the two ends of a through truss bridge (the bridge portals) and to transmit the wind loads on the bridge to the ground. A latticed portal is shown in Figure 10-1. Portal frames also were applied to buildings when it was found inconvenient to use diagonal brace rods. Although this technique applied more commonly to the transverse frames in a mill building, there are also cases where portal frames were applied to longitudinal frames. Examples of single bay portal frames are trusses provided with knee braces (see, for instance, Ketchum, 1903, p. 342, Transformer Building Section), whereas multiple bay portal frames can be seen in the longitudinal direction of many mill buildings (Ketchum, 1903, p. 368, ATSF RR locomotive shop). Portal frame structures also were used as bents in mill buildings. A combination of knee braces and moment connections between the truss and the columns, or the columns and the foundation were used to resist the lateral forces.

Milo Ketchum proposes an analytical treatment of portal frames of various configurations. This analysis depends in general on establishing equilibrium conditions in the frame and distributing the lateral force due to the wind to the supports of the portal. Some of the details of this analysis are illustrated in Figure 10-2. For many pinned base portal frames, one-half of the total horizontal force R is assumed to be distributed to each of the columns so that the horizontal reaction at the base of each column is $R/2$. The vertical reaction of the windward column is $-Rh/S$ (uplift) and at the leeward column is $+Rh/S$, where h is the height of the frame and S is the spacing of the columns.

Figure 10-1. Sayre, PA, converted railroad bridge over Cayuta Creek, circa 1910.

Internal forces in frames of various types can be worked out in general, without assuming a distribution of the total force to the base, according to the configuration of the frame. For the frame that Ketchum labels (a) (see Figure 10-3), the forces in the bars are found by successive sections based on the known external forces. The section cut X-X on Figure 10-2 shows that the force in GC = $-V \sec \theta$, by taking moments about the upper right corner of the frame. It is further found that the force in the three dashed bars is 0. The force in the other diagonal can be found from the section Y-Y to be $+V \sec \theta$. Similarly, based on section Y-Y, taking moments about D, it is found that the force in bar HG = $-Hd/(h - d)$. Finally, based on moments about C, the force in bar GF is found to be $[R(h - d) + Hd]/(h - d)$. Considering the wind in the direction shown only and considering the right side of the frame, GF is called the portal strut, GC the portal tie, and the remaining members the portal web. For the other side of the frame or for an opposite wind direction, the positions of the strut and tie are reversed. Whereas in the case shown the web members are 0 force, this is not true in general for other portal types, as is discussed following. The maximum shear in the posts is found to be equal to the force in HG, and the maximum moment in the posts is $M = Hd$.

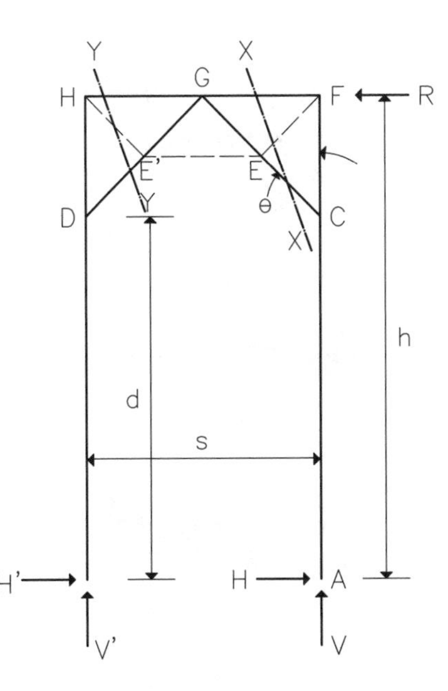

Figure 10-2. Type (b) portal frame shown to illustrate the nomenclature used by Ketchum and throughout this chapter to find the forces in the elements of a portal frame.
Source: Ketchum (1903).

In Figure 10-3, Ketchum (1903) presents six different types of portal frames, designated (a) through (f), for which he describes the determination of the design force or bending moment in each of the principal components: portal column (subject to axial force and bending moment), portal strut (subject to compression), portal tie (subject to tension), and portal web (subject to axial force resulting from shear). Figure 10-3 also identifies the principal components. C. L. Crandall (1888) also discusses type (e) and type (f) and reaches the same conclusions as Ketchum. To these may be added the types of portal frame discussed by Crandall, which are shown in Figure 10-4. He describes a lattice frame with a portal tie in a circular arc, type (g), and an arched lattice frame, type (h). Mansfield Merriman and Henry Sylvester Jacoby (1902, p. 273) discuss a composite portal frame with elements of type (e) and type (f) in a single portal frame. Portal types (a) through (h) are shown in Table 10-1, along with the formulas proposed for the determination of the stresses in the principal components of each portal frame. A variant of type (g) in which a straight lattice beam is placed on top of a pair of quarters circle braces is shown in the view in Figure 10-5 from the Harrisburg, PA, train depot. A sketch analysis of a portal frame of type (a) is available in the 1910 student notes of Norman Maddock. John Alexander Low Waddell (1884) includes a brief discussion and algebraic analysis of two types of portals (pp. 49–53), one a single-panel variant on type (c), the other type (f).

The forces in the principal members of each of the types of portals shown can be calculated, as described in Ketchum. These forces, which depend primarily on the geometry of the portal, can be compared by means of Table 10-1. See also Chapter 15 for a description of the graphical solution of these portals. As a representative example, consider the

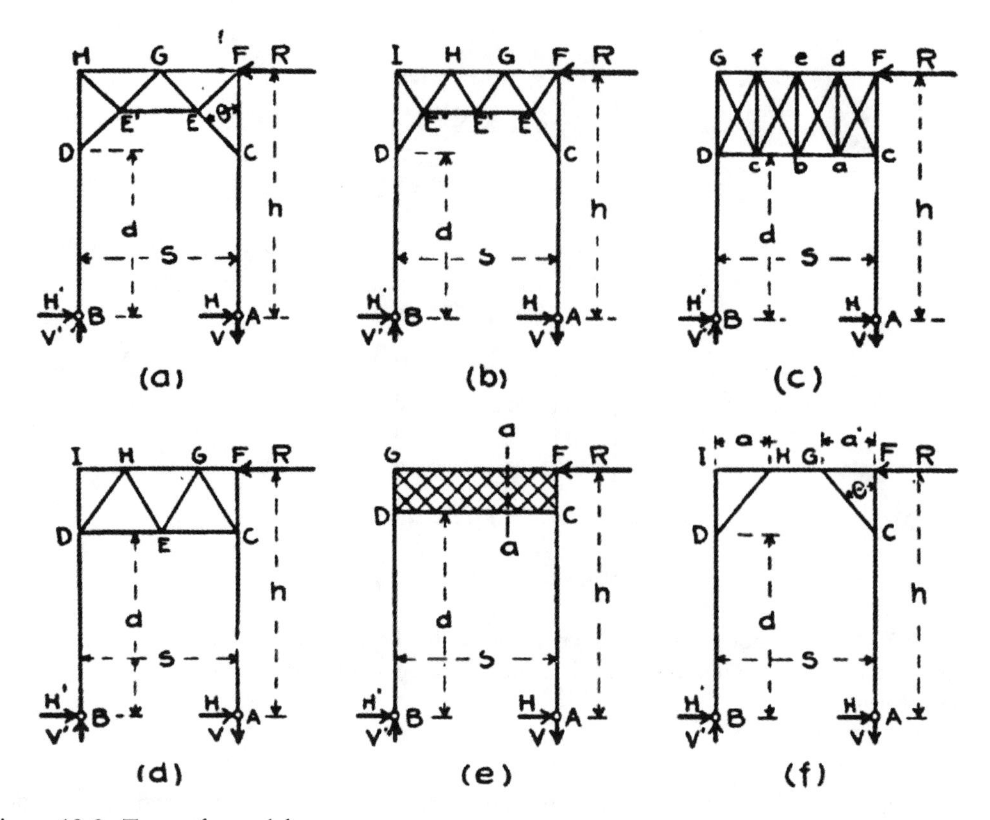

Figure 10-3. Types of portal frames.
Source: Ketchum (1903).

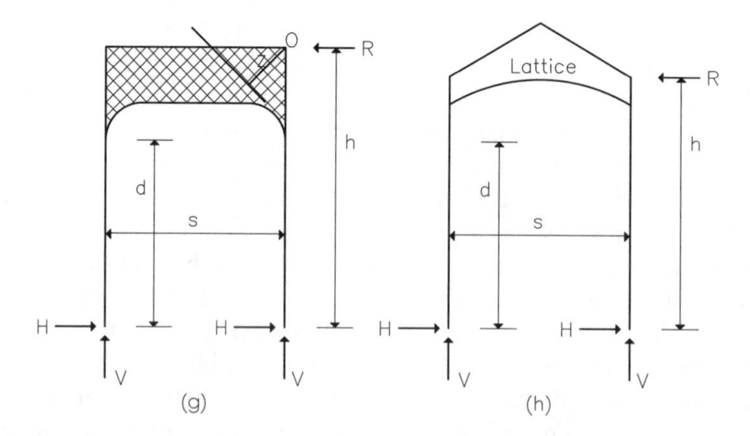

Figure 10-4. Additional types of portal frames (g) (left) and (h) (right).
Source: Crandall (1888).

calculation of the maximum web member force in configuration (b) (see Figure 10-6). The calculation is done by cutting a section through bars GF, EF, and EC (in a procedure described as Ritter's method in Chapter 7). The calculation of the force in Bar EF is accomplished by taking moments about G and finding the horizontal component of EF, that is, by

TABLE 10-1. *Forces in Portal Struts, Portal Ties, and Web Members in Various Portal Frames Shown in Figures 10-3 and 10-4*

Portal type	Portal strut force	Portal tie force	Maximum web member	Note
a	$R+H\frac{d}{h-d}$	$\frac{Hh}{h-d}\csc\theta$	0	
b	$R+H\left[\frac{(h+d)}{(h-d)}\right]$	$\frac{Hh}{(h-d)}\csc\theta$	$\frac{Hh}{(h-d)}\csc\theta$	E at midpoint of GC web force EF
c	$R+H\frac{d}{h-d}$	$\frac{Hh}{h-d}$	$\frac{2Hh}{s}\csc\theta$	tension only web
c′	$R+H\frac{d}{h-d}$	$\frac{Hh}{h-d}$	$\frac{Hh}{s}\csc\theta$	panel shear forces divided in half
d	$R+H\frac{d}{h-d}$	$\frac{Hh}{h-d}$	$\frac{2Hh}{s}\csc\theta$	
e	$R+H\frac{d}{h-d}$	$\frac{Hd}{h-d}$	$\frac{Hh}{2s}d \div s_l$	simplification, considering lattice bars resist shear only
f	$R+H\frac{d}{h-d}$	$\frac{Hd}{h-d}\csc\theta$	—	
g	$max\left[\left(Hh-\frac{2Hhx}{s}\right)\div y\right]$	horiz. component $\frac{Hd}{h-d}$	$2Hd/s$—horiz. component in tie at base	using Ketchum notation

θ is the vertical angle of the lattice bar closest to the end post, as shown in portal (a)
s_l is the spacing of the lattice bars (supposed at 45-degree angle)
x is the horizontal distance from the column to the point where forces are being evaluated
y is the overall vertical distance between the portal strut and the portal tie at x

transmitting the force in EF to act at a point along the line of action EF, directly below the center of moments G.

Moment equilibrium is written

$$(\mathrm{EF})\sin\theta(h-d)+Hh-\frac{V}{2}(h-d)\tan\theta=0$$

substituting *2Hh/s* for *V* and solving for force EF gives the formula shown in Table 10-1.

The cases (g) and (h) investigated by Crandall may require further explanation. In the case marked (g) in Figure 10-4, the actual calculation of the forces is relatively difficult, so he presents the strut force as the moment acting at a section divided by the vertical distance between the strut and tie and says that the maximum can be found "near enough after one or two trials (1888, p. 77)." Similarly, the tie force is calculated by taking external moments divided by the distance *z*, shown on the diagram in Figure 10-4, from the tangent to the curved part of the tie to the strut, for a center of moments designated O. According to Crandall, the critical center of moments can be safely taken as the intersection of the column and the portal strut. The horizontal component given in the Table 10-1 is resisted by the gusset plate at the end of the curved bottom chord. In the design marked (h), the strut, tie, and web shear force vary continuously and must be found by calculation of shears and

Figure 10-5. Portals at Harrisburg, PA, train station (HAER PA,22-HARBU,23–25).
Source: Photograph by F. Harlan Hambright.

moments at a section through the portal. The horizontal component of the flange force about a center of moments is (using Ketchum's notation as follows)

$$[H(h+u) - Vx - Hu] \div y$$

where x is the horizontal distance from the portal/column joint to the center of moments, y is the vertical distance between the top and bottom flange, and u is the rise of the top flange from the joint. Web shear is the total shear $2Hh/s$ less the vertical component in the bottom flange plus the vertical component in the top flange.

A similar treatment of similar portal types is available in A. Jay Du Bois (1896, pp. 309ff.) He deals algebraically with the case of overhead x-bracing and with the case of a knee brace for both a vertical and an inclined end post.

Crandall (1888) discusses further the effect of the inclination of the end post, which is common in truss bridges on the stresses, particularly in the portal strut and tie. Where this is done, there is no effect on the horizontal forces in the portal, whereas vertical force components increase by a factor of sec ϕ, where ϕ represents the angle of inclination of the portal from the vertical. He also elaborates on some of the particular issues in the determination of the stresses in a portal tie that is curved into a circular arc to increase the headroom

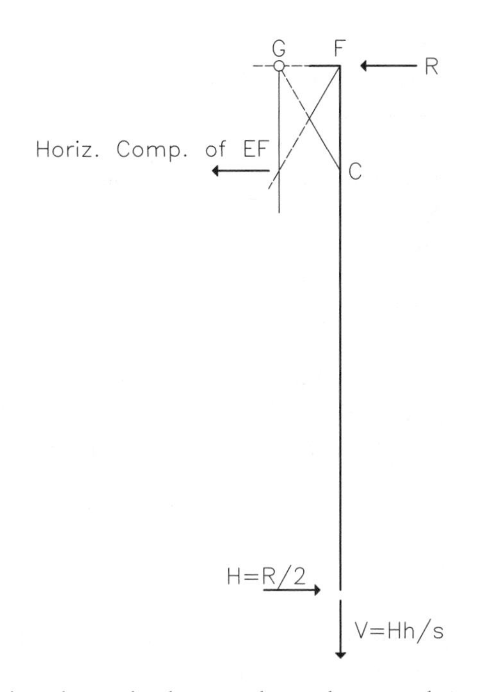

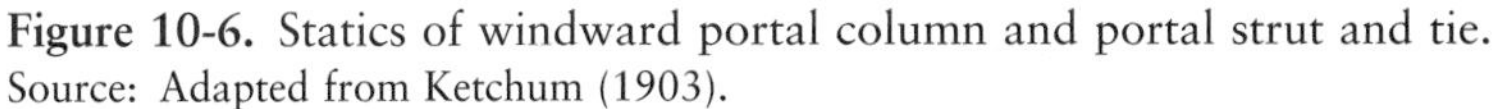
Figure 10-6. Statics of windward portal column and portal strut and tie.
Source: Adapted from Ketchum (1903).

at the entrance to the portal. The complete analysis of a portal, as described by Crandall, requires the determination of the wind loading on the entire truss. Crandall (1888, p. 5) proposes a loading of 150 lbs/ft as an adequate estimate of the total lateral load on the upper system. The load on the portal is this load multiplied by the span of the bridge and divided by 2. The load is calculated similarly for the lower lateral system.

When multiple portals are used, as along the length of a mill building, Ketchum shows two methods for finding the distribution of the horizontal force to the columns. The first is to distribute vertical reaction to the columns in proportion to the distance from the center of the multiple portals. This is the method now known as the "cantilever method." When the vertical column reactions are known, it is possible to find the horizontal forces in the columns. A much simpler procedure, preferred by Ketchum, is to divide the total horizontal force by the number of columns. This differs from the modern "portal method" only in assigning a equal shares of the horizontal load to the exterior and interior columns.

Similar treatments to Ketchum's and Crandall's are offered by other authors. Du Bois (1896) has a comprehensive article on wind bracing for bridges in which he gives a general approach to a bracing system consisting of an overhead strut, a horizontal tie at the clearance level of the portal, and diagonal brace rods. He considers this variant of type (d) for a bridge with inclined end posts (or "batter braces"), subjected to wind load. He further notes that the total wind load on a bridge is conventionally taken as 75 lbs/lineal ft. After finding similar moments in the batter braces to those determined from the equations in Table 10-1, he divides them into tension and compression based on the distance between the centroids of the channels making up the end post and asserts that the compressive stress must be added to the compressive stress that results from the loading on the bridge.

Three prevailing procedures for portal frame analysis for mill building bents are summarized in the bachelor's thesis by Arthur Shumway (1904). In this thesis, Shumway investigates three methods for the determination of portal frame stresses in mill buildings: an analytical method by James Greenleaf (1896) and graphical methods by Jerome Sondericker (1896) and by Ketchum (1903). Shumway's prototype mill building frame has knee braces and is shown in Figure 10-7.

Greenleaf presents a strictly analytical determination of the forces in the columns and knee braces of a portal frame. He subdivides the portal frames into classes A (columns pinned, no knee braces), B (columns fixed, no knee braces), and C (knee braces provided) and investigates individual classes of column fixity within each of these basic cases. The result is formulas for the main cases used in the construction of mill buildings. In a later study, Shumway omits the cases without knee braces and repeats and amplifies Greenleaf's calculations, using an example of a mill building with a fan truss roof. The compilation of stresses of both Greenleaf (case B-4) and Shumway (case 1-d) for the case of a mill building bent with knee braces, pinned at top and bottom (Figure 10-8) is as follows:

$C = A(h - d)/d$
$B = A + C$
Above point b, the maximum bending moment is at *b* and equals *Cd*.
$M_x = Cx$ (*x* is the distance from the top of the column)
Below point b, the maximum bending moment is at *b* and equals *Cd*.
$M_x = Cx\text{-}B(x - d)$
$V_a = 1/s(\Sigma W\ 1/2l)$

where

V_a is the vertical reaction at the column;
C designates the force in the column top;

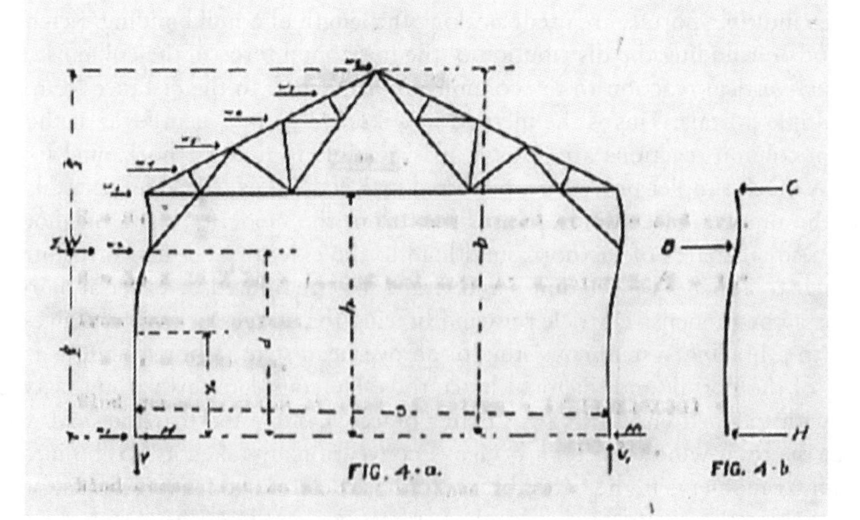

Figure 10-7. Shumway's typical mill building bent.
Source: Shumway (1904).

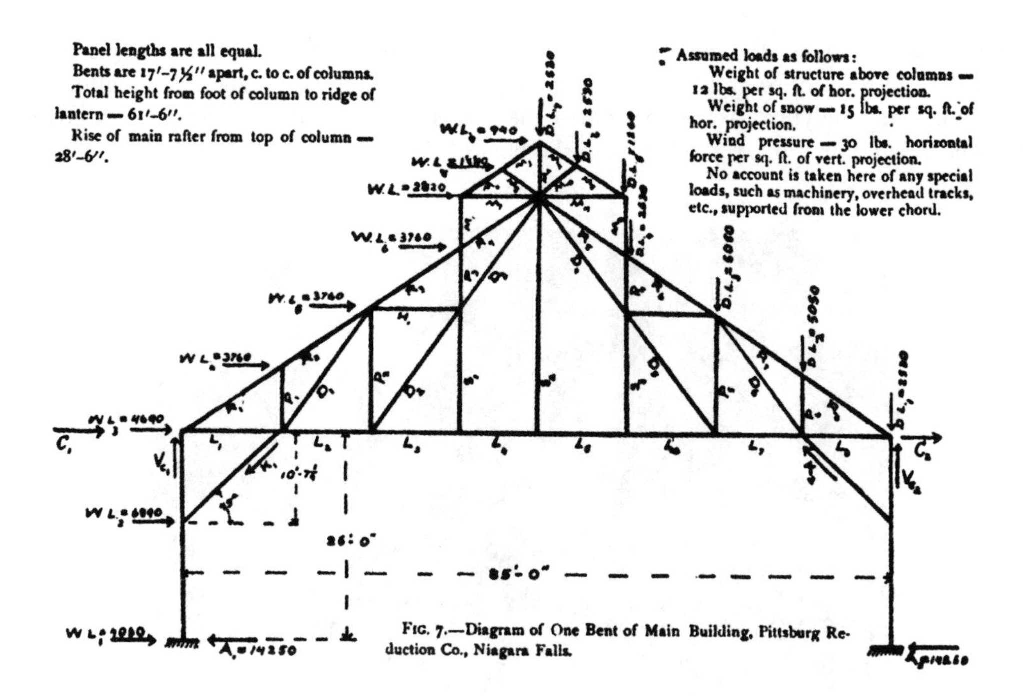

Figure 10-8. Greenleaf's typical mill building bent.
Source: Greenleaf (1896).

TABLE 10-2. *Similar Building Case Investigated by Greenleaf (1896) and Shumway (1904)*

Characteristic	Greenleaf (case B-4)	Shumway (case 1-d)
l overall height	61 ft 6 in.	30 ft
h height of columns	26 ft	20 ft
d moment arm of knee brace	10 ft 7¼ in.	5 ft
s spacing of columns	17 ft 7½ in.	16 ft
A reaction at base of column	14,250 lb	5400 lb
K stress (force) in knee brace	39,680 lb	30,540 lb
C force in column top	20,700 lb	16,200 lb

B is the horizontal force at b, the point of attachment of the knee brace to the column;
A is the horizontal reaction at the bottom of the column;
h is the height of the column;
d is the distance from point b to the top of the column;
l is the height of the ridge; and
s is the span of the roof.

Shumway and Greenleaf investigate mill building bents with different characteristics according to these rules and determine the forces shown in Table 10-2. The notation used is that of Greenleaf. The table showing all of the cases of Greenleaf's investigation of the bent shown in Figure 10-8 is displayed in Table 10-3. An application of knee braces to the truss column bents of this type in a nineteenth century mill building is shown in Figure 10-9. Following the analytical determination of the forces in the column, Greenleaf proposes a graphical determination of the forces in the bars of the truss.

Figure 10-9. Knee braces in a beam-column connection in a mill building bent, Whiting Foundry Equipment Company, Harvey, IL (1893): (a) general view, (b) detail of knee braces.
Source: Photographs by Michael Powers.

TABLE 10-3. *Greenleaf's (1896) Compilation of Stresses in Mill Building Bents with Various Support Conditions*

COMPILATION OF STRESSES FOR THE FRAME SHOWN IN FIGURE 7.

The stresses are given for all the various conditions as to the use or absence of knee braces and gusset plates, and the different assumptions regarding the connections at the ends of the columns. The data upon which the calculations are based are given on page . + denotes tension and — denotes compression. All forces are given in pounds.

Designation of Member of Truss, Bending Moment, Etc.	Stresses Due to Dead Load.	Case A—1.		Case A—2.		Case A—3.		Case B—1.		Case B—2.		Case B—3.		Case B—4.		Case C—1.		Case C—2.	
		No knee braces Cols. pinned at top. Cols. fixed at base.		No. knee bra. Cols. fixed at top. Cols. pinned at base.		No. knee bra. Cols. fixed at top. Cols. fixed at base.		Knee braces (2 L's.). Cols. fix'd at top. Cols. fixed at base.		Knee braces (2 L's.). Cols. pinned at top. Cols. fixed at base.		Knee braces (2 L's.). Cols. fixed at top. Cols. pinned at base		Knee braces (2 L's.). Cols. pinned at top. Cols. pinned at base.		Cols. stfn'd. by gussets from ft. of bra. up. Cols. pinned at top. Cols fix'd at ba.		Cols. stiffened by gussets from ft. of brace up. Cols. pinned at top. Cols. pinned at base.	
		Wind Load.	Max. + or —	Wind Load.	Max. + or —	Wind Load.	Max. + or —	Wind Load.	Max. + or —	Wind Load.	Max. + or —	Wind Load.	Max. + or —	Wind Load.	Max. + or —	Wind Load.	Max. + or —	Wind Load.	Max. + or —
Bend. Moment at top of Column (ft. lbs)		Zero	Zero	334100	334100	167100	167100	46800	46800	Zero	Zero	109700	109700	Zero	Zero	Zero	Zero	Zero	Zero
Bend. Moment at Base of Column (ft. lbs.)		334100	334100	Zero	Zero	167100	167100	125900	125900	130300	130300	Zero	Zero	Zero	Zero	109700	109700	Zero	Zero
External Force $C_1 = C_2$ (hor. at top of Column),		12850	12850	12850	12850	12850	12850	13250	13250	8410	8410	31050	31050	20700	20700	10350	10350	20700	20700
External Force K_1 (Stress in Knee Brace),		....	....			....	....	29140	29140	22300	22300	54310	54310	39680	39680	25040	25040	39680	39680
External Force K_2 (Stress in Knee Brace),		....	....	....	....	....	....	38890	38890	32040	32040	64050	64050	49420	49420	34780	34780	49420	49420
External Force Vc_1 (Vert. at left end of truss),		3940		3940		3940		10650		7020		24000		16240		8420		16240	
External Force Vc_2 (Vert. at right end of truss),		3940		3940		3940		17540		13910		30890		23130		15310		23130	
R_1 (Rafter, Left Hand),	−31800	+ 7100	−31800	Same as Case A—1.		Same as Case A—1.		−19100	−50900	−12600	−44400	−43200	−75000	−29100	−60900	−15100	−46900	Same as Case B—4.	
R_2 " " "	−31800	+ 2500	−31800					−23700	−55500	−17100	−48900	−47700	−79500	−33700	−65500	−19600	−51400		
R_3 " " "	−31800	— 800	−32600					— 8700	−40500	— 6400	−38200	−16600	−48400	−11700	−43500	— 6900	−38700		
R_4 " " "	−31800	— 5400	−37200					−13200	−45000	−10900	−42700	−21300	−53100	−14100	−45900	−11500	−43300		
R_5 " Right "	−31800	— 7100	−38900					+31600	−31800	+25000	−31800	+55600	−31800 +23800	+41500	−31800 + 9700	+27600	−31800		
R_6 " " "	−31800	— 7100	−38900					+ 6800	−31800	+ 4700	−31800	+15100	−31800	+10100	−31800	+ 5400	−31800		
L_1 (Lower Chord, Left Hand),	+26400	— 1800	+26400					— 2000	+26400	— 2500	+26400	+ 200	+26600	+ 1200	+26400	— 1600	+22600		
L_2 " " " "	+22600	— 3700	+22600					+ 4000	+22600	— 500	+22600	+ 8000	+30600	+ 4000	+26600	+ 200	+15300		
L_3 to $_6$ " " " "	+15100	— 7000	+15100					— 9700	+15100	— 1700	+15100	— 9700	+15100	— 9700	+15100	— 9700	+ 5400		
L_8 " " Right "	+26400	— 7000	+26400					−13900	+26400	— 2400	+26400	−26900	— 500 +26400	−13900	+26400	−12600	+26400		
L_7 " " " "	+22600	— 7000	+22600					−22700	— 100 +22600	— 8000	+22600	−15200	+22600	−22700	— 100 +22600	−18800	+22600		
D_1 (Diagonal, Left Hand),	+ 6300	+ 3200	+ 9500					+38100	+44400	+ 2900	+29200	+51200	+57500	+38100	+44400	+25200	+31500		
D_2 " " "	+12500	+ 5500	+18000					+22900	+35400	+15300	+27800	+29700	+42200	+22900	+35400	+16500	+29000		
D_6 " Right "	+ 6300	Zero	+ 6300					−43500	−37200 + 6300	−28300	−22000 + 6300	−56400	−50100 + 6300	−43500	−37200 + 6300	−30800	−24500 + 6300		
D_5 " " "	+ 12500	Zero	+12500					−21700	— 9200 + 12500	−14100	— 1600 +12500	−28600	−16100 +12500	−21800	— 9300 +12500	−15400	— 2900 +12500		
P_2 (Vertical, Left Hand),	−10100	— 4400	−14500					−18300	+28400	−12300	−22400	−23700	−33800	−18300	−28400	−13200	−21300		
P_s (Vertical, Right Hand),	−10100	Zero	−10100					+17400	−10100 + 7300	+11300	−10100 + [illegible]	+22900	−10100 12800	+17400	−10100 + 7100	+19400	−10100 [illegible]		

252

Portal frames were a widely used, statically indeterminate structure type in the late nineteenth century and beyond. The methods generally available for their analysis used various forms of approximations to arrive at a reasonable result. The approximations included the use of approximate values of wind loading, such as a wind load per lineal foot on a bridge superstructure, and the approximate division of lateral loads on a frame between the two columns. Once the loads and reactions were known, analytical methods for the analysis of portal frames relied on taking successive sections and taking moments about strategic points, effectively Ritter's method for trusses, described in Chapter 7. The forces in the bars in portal frames could also be determined graphically by the methods that are described in Chapter 15.

References Cited

Crandall, C. L. (1888). *Notes on bridge stresses and bridge designing for use in the civil engineering department of Cornell University*. Mimeograph print.

Du Bois, A. J. (1896). *The stresses in framed structures*, 10th Ed. John Wiley and Sons, New York.

Greenleaf, J. L. (1896). "The treatment for wind pressure in mill construction." *School of Mines Quarterly*. Columbia University, Vol. 17, November 1895–July 1896.

Ketchum, M. (1903). *Design of steel mill buildings*. Engineering News Publishing, New York.

Maddock, N. (1910). Civil Engineering Notes. Collection MSVF/968-0030, Penn State University Libraries Special Collections.

Merriman, M., and Jacoby, H. S. (1902). *A text-book on roofs and bridges. part 3, bridge design*, 4th Ed. John Wiley and Sons, New York.

Shumway, A. (1904). *Wind stresses in a steel frame mill building*. Bachelor's thesis, Cornell University, Ithaca, New York.

Sondericker, J. (1896). *Notes on graphic statics*. J.S. Cushing, Boston.

Waddell, J. A. L. (1884). *The designing of ordinary iron highway bridges*, 5th Ed. John Wiley and Sons, New York.

Part III

Graphical Methods of Analysis

11

Introduction to Graphical Methods of Analysis

Graphic statics is the representation of known forces and the solution of unknown forces using scaled diagrams. To display all of the properties of a plane force, two diagrams are necessary. In one diagram, drawn to a length scale, the direction, location, and sense, positive or negative, of a force are represented, whereas in the other, which is drawn to a force scale, the magnitude, direction, and sense of a force are shown. Two forces and their equilibrant (shown dashed) are drawn in the two complementary diagrams in Figure 11-1. Forces combine in the force diagram by placing the vectors representing their magnitude and direction in a tip-to-tail configuration. The sum of the three forces in the diagram, including the equilibrant force, is 0—the system of forces forms a closed loop with the same point of beginning and ending. In the funicular diagram, the magnitude of the force is not represented. The resultant force or its opposite—the equilibrant force—is located so that its line of action passes through the intersection of the lines of action of its component forces.

To find the resultant of a larger system of forces, as in Figure 11-2, an arbitrary point (known as the pole) is chosen in the force diagram, and the forces in the diagram are resolved into components passing through the pole. These components are then plotted in their correct direction and location on the funicular diagram, with the result that the location of the equilibrant or the resultant is known. This is demonstrated in the diagrams in Figure 11-2. Forces *ab*, *bc*, *cd*, and *de* are shown in the force diagram and the directions of these forces in the funicular diagram. Given a pole O in the force diagram, the components of *ab* are O*a* and O*b*, labeld *a* or *b*, and the components of the resultant are O*a* and O*e*. Thus, if O*a* and O*e* can be found on the funicular diagram, the resultant, whose line of action passes through the intersection of *Oa* and *Oe*, can be located. Each pair of resultants of known forces is laid out in such a way that the two components intersect on the line of action of the force *ab*, *bc*, *cd*, or *de*. In this way the lines of action of the resultants *Oa* and *Oe* are located on the diagram, and the resultant can be drawn passing through the intersection of forces *Oa* and *Oe*. The strings *a*, *e*, and the resultant in the force and funicular polygon reduce to a triangle like that described in Figure 11-1. Effectively, any two strings of the funicular polygon intersect in a point on the line of action of the resultant of the forces between these two strings. Thus, for instance, string *b* and string *e* on the funicular polygon in Figure 11-2 intersect in a point on the line of action of the resultant of forces *bc*, *cd*, and *de*. This construction has many uses, a few of which will be described in detail in the following chapters.

The relationship between projective geometry and graphic statics, which is described more thoroughly in Box 11-1, is noted by Karl Culmann (1875) and discussed extensively by A. Jay Du Bois (1877), Henry T. Eddy (1878), and others. Culmann's program appears to have been to transform statical calculations to graphic language. This is evident in his book on the topic, *Die Graphische Statik*. In this work, he provides methods for the graphical calculation of centers of gravity, moments of inertia, beams, and frames. He investigates the properties of forces in space through graphical diagrams. Erhard Scholz (1989) gives a modern account of Culmann's attempts to bring out the projective geometry character of

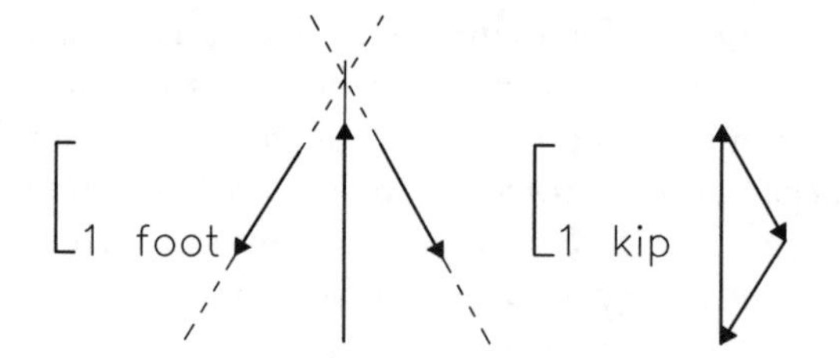

Figure 11-1. Simple spatial and force diagrams for three concurrent forces in equilibrium.

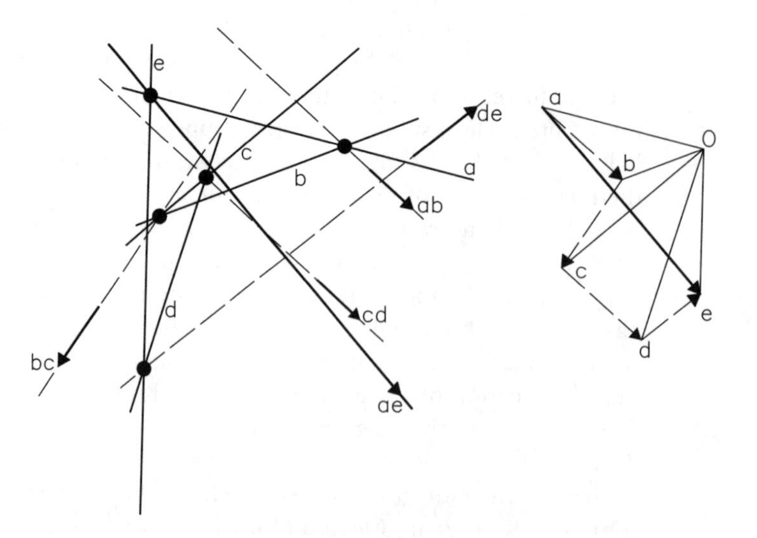

Figure 11-2. Force and funicular polygons for a system of five plane forces.

Box 11-1. Projective Geometry and Graphic Statics

For a more formal treatment of projective geometry, the reader is referred to H. S. M. Coxeter (1955). The basis of plane projective geometry is the acceptance of an ideal point as the point of intersection of parallel lines, located in the direction of the parallel lines "at infinity." Lines close to parallel can be seen to intersect at increasingly remote distances as the angle between the lines becomes smaller, with "intersecting at infinity" considered as the limiting case. Two points at infinity can be used to define the line at infinity, whereas an ordinary point and a point at infinity define a line in a given direction.

One of the results of this conception of plane geometry is that Euclid's first axiom, "exactly one line can be drawn through two points" is interchangeable with the modified fifth axiom that any two lines intersect in exactly one point. So, any statement regarding lines and points and their incidence can be transformed to the "dual" statement, simply by interchanging "points" and "lines." Following are examples of this dualization:

> The dual figure to three collinear points (three points incident with a line) are three concurrent lines (three lines incident with a point).
>
> The dual figure to any polygon is another polygon with edges and vertices interchanged. Each vertex in a pentagon, for instance is incident with two edges, and each edge in the dual figure is incident with two vertices.

Two constructions are particularly important in projective geometry: the construction of Desargues's theorem, discussed in Box 11-2, and the construction of harmonic points on a line, discussed in Box 11-3.

To add to the construction of the force and funicular polygon described in Chapter 11, one can choose an arbitrary point, called the pole of the force polygon, and use it to construct the components of each of the forces. These force directions are placed on the funicular polygon in such a way that they intersect on the line of action of each of the three forces in the system (Figure B11-1-1).

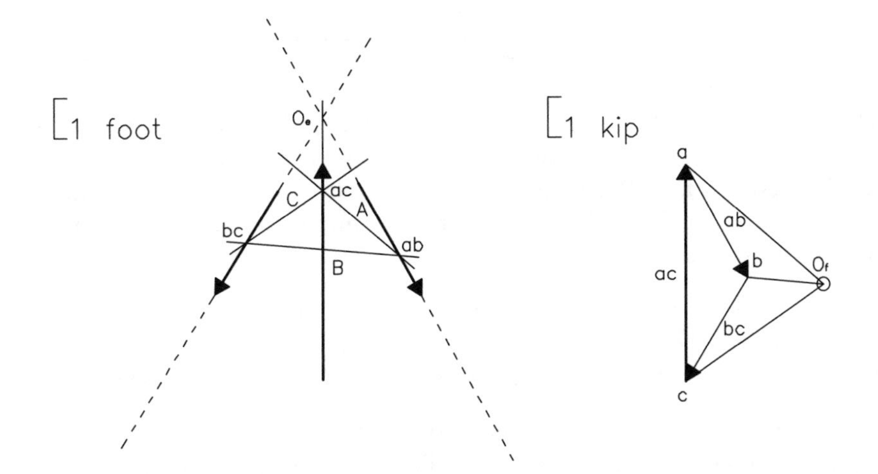

Figure B11-1-1. Construction of force and funicular polygon for three concurrent forces.

This construction shows the *reciprocal* nature of the two triangles. The triangle *ac ab, bc* on the force diagram represents three forces in equilibrium (and the corresponding concurrent lines on the funicular diagram), and the *funicular* triangle *abc* (and the corresponding concurrent lines on the force diagram) represents the lines of action of a set of internal forces, or strings, in equilibrium with the applied loads.

The connection to projective geometry, as described, can be seen in the duality between the two diagrams shown in Figure B11-1-1: edges of the force polygon correspond to vertices of the funicular polygon and *vice versa*. The three edges of the force polygon correspond to three concurrent lines on the funicular polygon and *vice versa*. This is the configuration shown in Figure B11-1-1 and Table B11-1-1; however, the diagrams do not represent a pair of dual figures. In this configuration, the force polygon consists of six lines and four points: each line is incident with two points, and each point is incident with three lines. The funicular polygon consists of six lines and four points: each line is incident with two points and each point is incident with three lines. The dual figure of each of these polygons would have four lines and six points, each point incident with two lines and each line incident with three points. Thus, the dual figure would be a complete quadrilateral as described later in Box 11-3.

As Culmann points out (1875, pp. 280–282), extending the lines *ABC* of the funicular polygon or *ab*, *bc*, *ca* of the force polygon to their intersection with the line at infinity and adding the line at infinity results in two configurations of seven lines and seven points, with three points incident with each line and three lines incident with each point, that is a self-dual configuration. The roles of each of the elements in the force and funicular polygon are reciprocal, as described in Table B11-1-1.

TABLE B11-1-1. *Reciprocity of Force and Funicular Polygons Following Culmann (1875)*

Feature	Force	Funicular
External forces	External force lines *ab, bc, ca*	Points of application *ab, bc, ca*
Internal forces	Points of intersection *a, b, c*	Internal force lines *a,b,c*
Pole	O_f	Line at ∞
Pole	Line at ∞	O_e

the graphical calculations that he developed in *Die Graphische Statik*. Du Bois, in the preface to his book on graphic statics, quotes Jean-Victor Poncelet to describe the developments in geometry that resulted from the work of mathematicians and claims that projective geometry, or geometry of position will eventually replace metric, or Euclidean, geometry, in the vocabulary of the engineer:

> Little by little, algebraic knowledge will become less indispensable, and science, reduced to what it needs to be, to what it should already be, will thus be placed within the grasp of that class of men who have but a few rare moments to dedicate to its study. (p. xlviii)

In fact, throughout the rest of his book on graphic statics, Du Bois's use of projective geometry is very limited. He discusses the application of Desargues's theorem (discussed in Box 11-2) at one point in the development of the graphical calculation of support moments

Box 11-2. Desargues's Theorem

According to Desargues's theorem, if two triangles have their vertices on lines incident with a common point, then the points of intersection of the corresponding sides of the triangle lie in a common line. The funicular polygon in Figure B11-2-1 is an example of a triangle with its vertices incident with a common point. To see how Desargues's theorem works, one can add another funicular triangle *a'b'c'* to the diagram in Figure B11-2-1. Then the intersection of sides *a* and *a'*, *b* and *b'*, and *c* and *c'* all lie in a common line, which is labeled the Desargues line on the diagram. The theorem holds equally for pairs of triangles whose vertices lie in parallel lines, that is, lines radiating from a common point at infinity.

In constructing the corresponding set of reciprocal lines *a'*, *b'*, and *c'* in the force polygon, it is found that the pole shifts in a direction identical to the direction of the Desargues line, as shown in Figure B11-2-2.

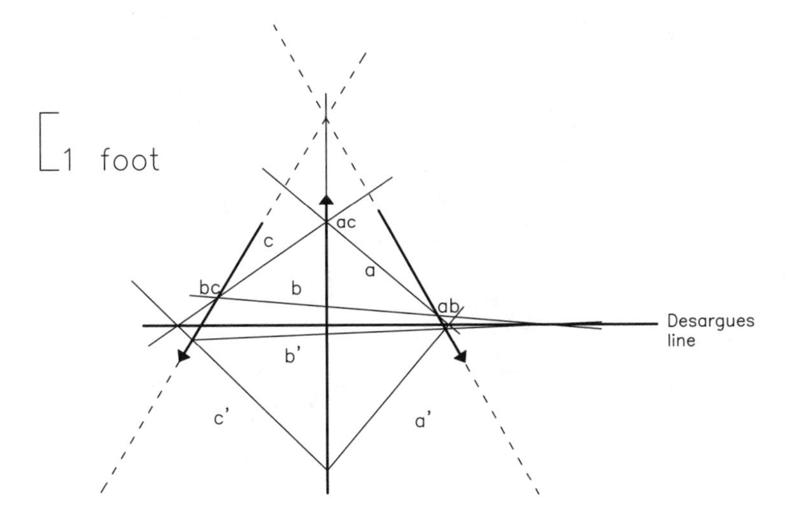

Figure B11-2-1. Desargues's theorem applied to funicular diagram for three concurrent forces.

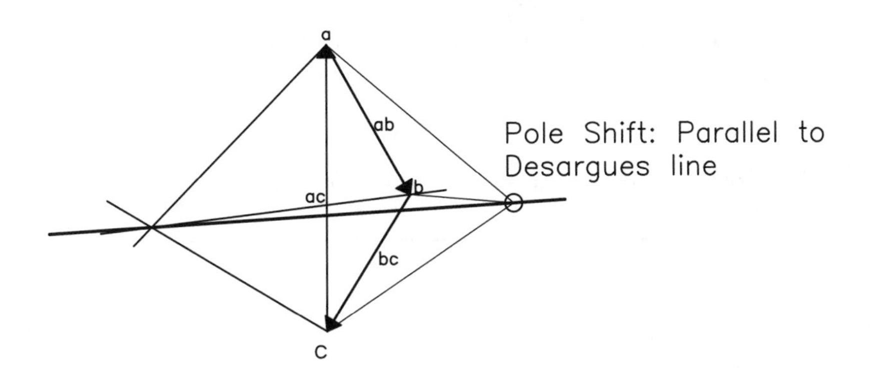

Figure B11-2-2. Force polygon for three concurrent forces with two alternative pole locations.

These facts about the force and the funicular polygon, and particularly about changes in the location of the pole in the force polygon, are illustrated in Eddy's diagram (Figure B11-2-3) in his article on reciprocity in graphic statics (1878). Eddy's diagram shows four forces *ab, bc, cd, de*, and their resultant *ae*, plotted on two different force polygons, with poles at *p* and *p′* and on two different funicular polygons. Because the resultant and any two reactions form a triad of forces in equilibrium, a consequence of Desargues's theorem is that the strings of the funicular polygon intersect on a common line, which will be called a Desargues line. The corresponding Desargues line is shown on Eddy's construction as *pp′* slightly inclined from the horizontal.

The main value of these methods lies in the applications to solving statics problems, and the graphic solutions of beam, truss, and arch problems is shown in following chapters.

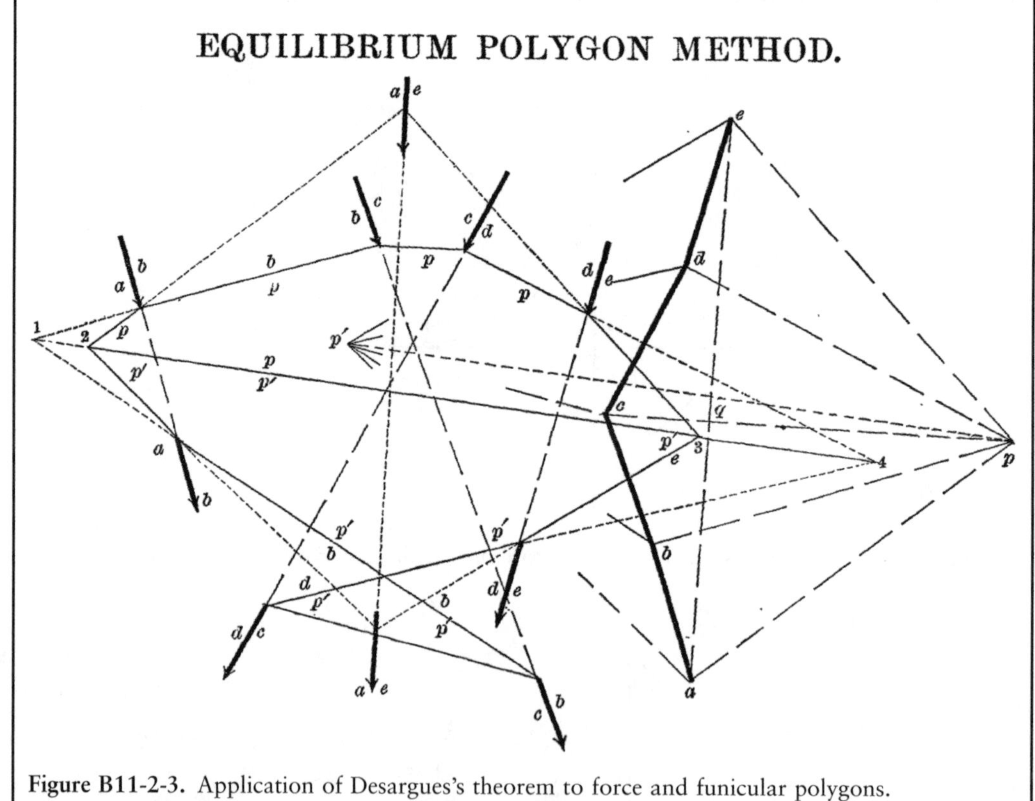

Figure B11-2-3. Application of Desargues's theorem to force and funicular polygons.
Source: Eddy (1878).

Box 11-3. Projective Geometry Transformations Following Coxeter (1955)

The study of projective geometry includes the study of the effect of transformations, such as perspectivities, and composition of perspectivities, known as projectivities, along with two-dimensional transformations of points to points and lines to lines (collineation) or points to lines and lines to points (correlation). These transformations generally do not preserve metric properties (distances) but do preserve the harmonic property described following. The two-dimensional transformations also preserve incidences, that is, for a collineation, three points incident with a given line remain so after transformation. For a correlation, the three points incident with a line transform to three lines incident with a point.

A quadrilateral with its two opposite sides extended to their point of intersection (that may be at infinity) is known as a "complete quadrilateral." By connecting the two points of intersection of pairs of opposite sides and then drawing the diagonals of the quadrilateral, the line is divided into four parts that respect a ratio called the "harmonic ratio."

The six points *ABCD* and *PQ* shown in Figure B11-3-1 form a complete quadrilateral. There is said to be a harmonic relationship among *P*, *Q*, *R*, and *S*. The point *S* is considered to be harmonic to *R* with respect to *PQ*, or reciprocally *R* is harmonic to *S* with respect to *PQ*, or *P* is harmonic to *Q* with respect to *RS*, or *Q* is harmonic to *P* with respect to *RS*. The first of these relationships may be abbreviated H (*PQ*, *RS*). This means that the line segments observe the harmonic ratio. Note that if *R* is at the midpoint between *P* and *Q*, *S* is at infinity.

The basic one-dimensional (linear) transformations of projective geometry are perspectivities and projectivities. A projectivity is a general transformation of a sequence of

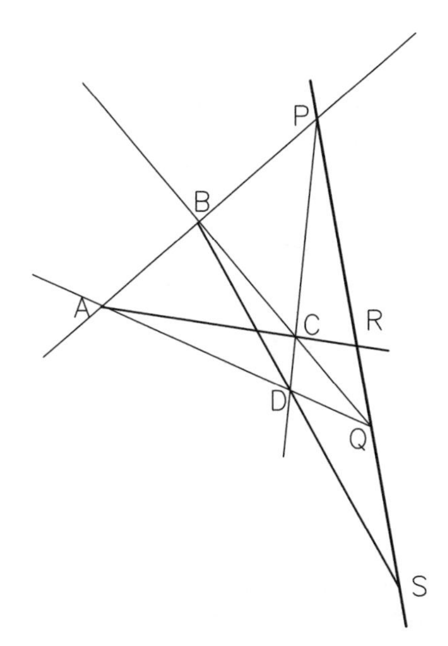

Figure B11-3-1. Complete quadrilateral ABCDPQ and harmonic construction H(PQ,RS).

points that preserves the harmonic property, whereas a projectivity is a projection of the points of a line onto points of another line. A perspectivity from line *a* to line *b* about a focus O is shown in Figure B11-3-2. It can be proven that perspectivities preserve the harmonic relationship, that is, if *H*(*AB*,*CD*), then *H*(*A'B'*,*C'D'*) and that every projectivity is a composition of perspectivities.

Two-dimensional transformations are more complex, but one—the homology—is of interest in graphic statics and is revisited in Chapter 13 in the section on the graphical analysis of arches. A homology is a transformation of a plane by an axis and a center. The transformation is determined by the center, the axis, and two pairs of transformed points. As shown in Figure B11-3-3, a homology is defined with the center O, axis *m*, and a single pair of transformed points *A*, *A'*. Because the axis consists of invariant points, the line *AB* intersects the axis *m* at an invariant point, so the line *A' B'* must intersect the axis at the same point, giving the location of *B'*. The transformation preserves incidence, that is, the transform of line *AB* passes through *A'* and *B'*.

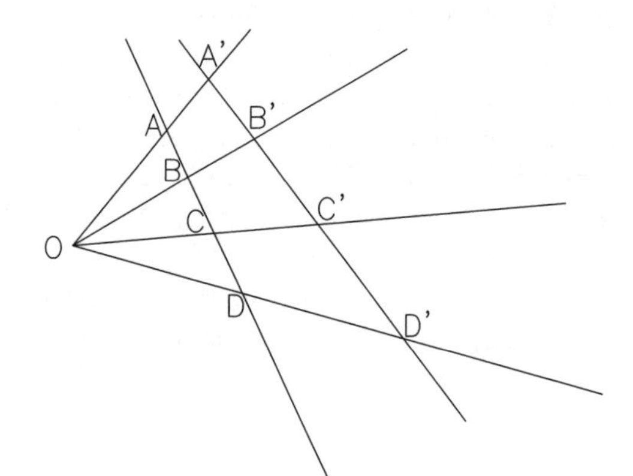

Figure B11-3-2. A perspectivity with center O.

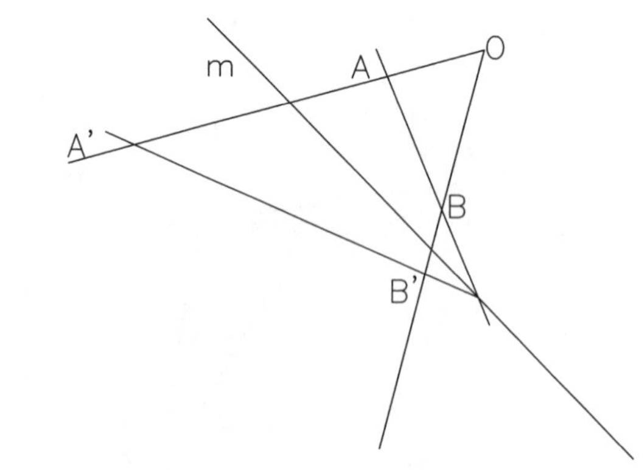

Figure B11-3-3. A homology with center O and axis *m*.

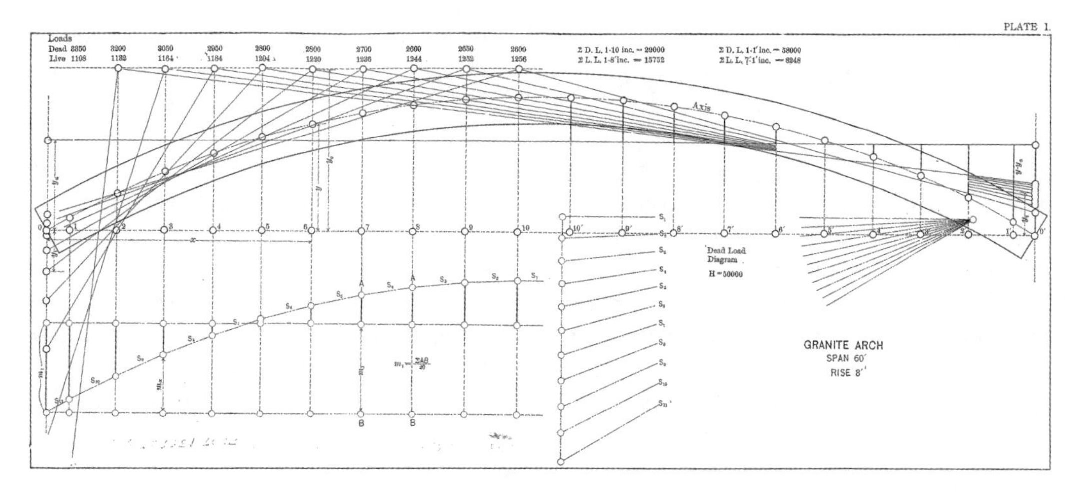

Figure 11-3. Graphic analysis of an arch force polygon is in lower right, illustrated by strings *S*. Funicular polygons are drawn superimposed on the diagram of the arch.
Source: Howe (1914).

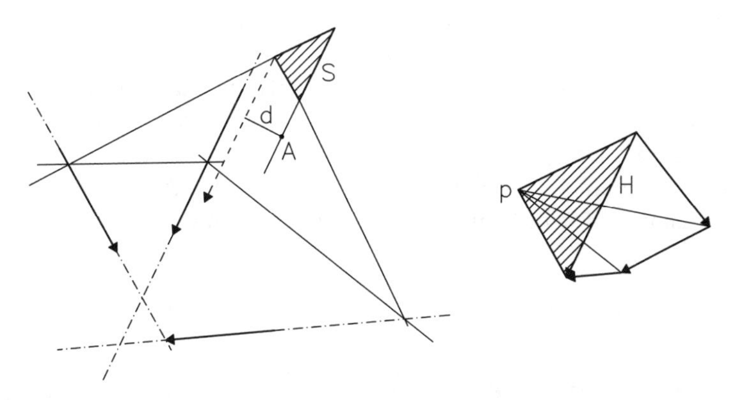

Figure 11-4. Culmann's Theorem: The moment of the resultant force about *A* can be calculated as the product of the force *H* and the length *S* instead of the resultant force times the distance *d*.

in a three-span beam. Other authors, such as Malverd Howe (1914) also provide an extensive graphical treatment of the topic of arch analysis (see Figure 11-3).

Culmann's Theorem

Culmann's Theorem is a result that will be invoked in the graphical analysis of beams and is also applicable to the analysis of arches and trusses. Culmann's Theorem is based on the similarity relationships between the force polygon and the funicular polygon. It allows the calculation of internal bending moments in beams on the basis of an evident force quantity in the force polygon and a length in the funicular polygon. Referring to Figure 11-4, given a system of forces shown on the funicular polygon and the corresponding force polygon, the moment of the resultant force about an arbitrary point *A* is equal to the magnitude of the resultant (shown on the force polygon) times the perpendicular distance from the line of action of the resultant to *A*, designated *d* on the funicular polygon. According to Culmann's theorem, this is equal to the perpendicular distance from the pole to the resultant of the force diagram, designated *H*, and the distance *S* on the funicular polygon, measured between strings representing the components of the resultant along a line parallel to the resultant passing through the point *A*. The theorem is a consequence of the similarity of the shaded triangles in the force and funicular polygong.

The widespread applications of graphical statics to structural design in the late 1800s are more easily seen through the investigation of examples of areas where these techniques were commonly used. In the following four chapters, graphical statics is examined in its application to three important structural types. First, the uses of these techniques for the investigation of the forces in the bars of trusses will be investigated—this is undoubtedly the most widespread and the most enduring of the applications of graphical statics. The uses of graphical statics in the analysis of arches then is investigated. The analysis of arches will be followed by the presentation of some particulars on the graphical analysis of beams and portal frames.

Although further examples of graphical constructions are provided in the following chapters, it is worthwhile to consider some of the possible uses of graphical statics in structural engineering. The principal uses of graphical statics in structural engineering are in the analysis of trusses, arches, and beams, which are covered in Chapters 12, 13, and 14. Figure 11-3 shows a graphical analysis of a low-rise symmetrical arch. The figure is taken from Howe (1914). Figure 11-5, taken from Charles Ezra Greene (1877), depicts the force polygon and the funicular polygon for a beam, a construction that will be examined in further detail in Chapter 14. In Figure 11-6, taken from Howard Drysdale Hess (1915), a

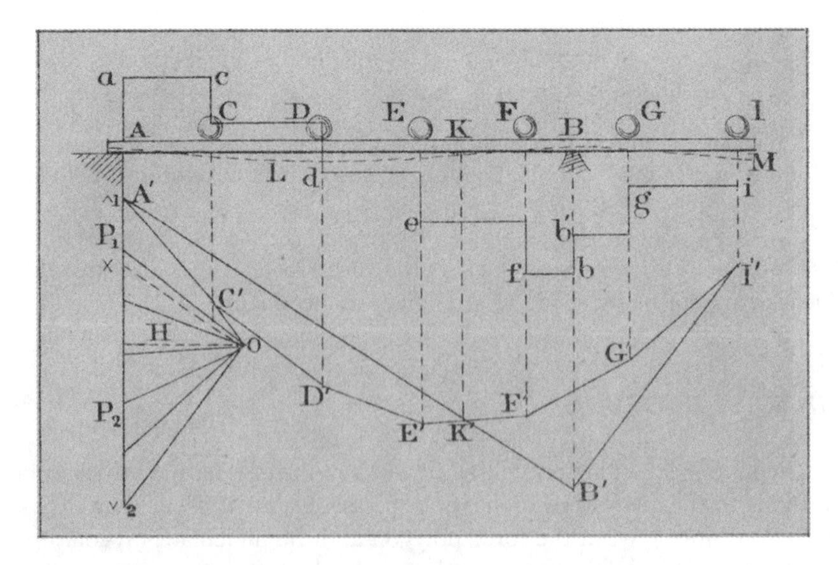

Figure 11-5. Illustration of the force polygon (O12) and the funicular polygon *A′C′D′E′K′F′G′I′* for a beam.
Source: Greene (1877).

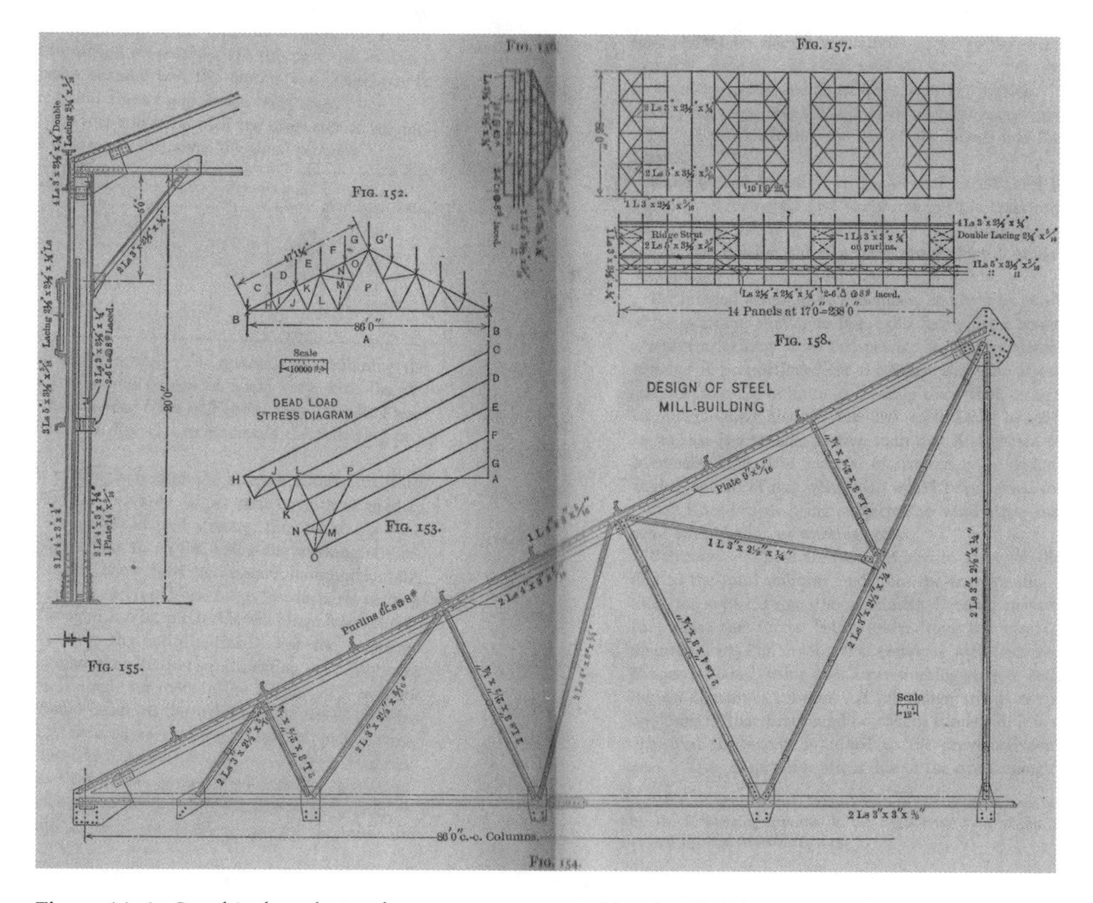

Figure 11-6. Graphical analysis of a truss, accompanied by detailed drawings of the truss.
Source: Hess (1915).

truss is shown accompanied by its graphical analysis and construction details. The topic of the graphical analysis of arches will be discussed in further detail in Chapter 13.

References Cited

Coxeter, H. S. M. (1955). *The real projective plane*, 2nd Ed. Cambridge University Press, MA.

Culmann, K. (1875). *Die graphische statik*, 2nd Ed. von Meyer and Zeller, Zurich (in German).

Du Bois, A. J. (1877). *Elements of graphical statics*, 2nd Ed. John Wiley and Sons, New York.

Eddy, H. T. (1878). On two general reciprocal methods in graphical statics. *Am J Math*, 1(4), 322–335.

Greene, C. E. (1877). *Bridge trusses; graphical method*. John Wiley and Sons, New York.

Hess, H. D. (1915). *Graphics and structural design*, 2nd Ed. John Wiley and Sons, New York.

Howe, M. A. (1914). *Symmetrical masonry arches*, 2nd Ed. John Wiley and Sons, New York.

Scholz, E. (1989). *Symmetrie, gruppe, dualität. zur beziehung zwischen theoretische mathematik und anwendungen in kristallographie und baustatik des 19. jahrhunderts*. Birkhäuser Verlag, Basel, Boston (in German).

12

Graphical Analysis of Trusses

Like analytical methods, graphical analysis methods for trusses are divided into the method of joints and method of moments (now known as method of sections). The graphical method of joints was commonly used in trusses for buildings, due to its simplicity, ease of application, and self-correcting characteristics. The use of this method persisted through the 1950s (Sahag 1958, Turner 1966). The graphical method of moments was less commonly used: it is less straightforward and, like its analytical counterpart, only determines the forces in the three members cut in a given section. The two procedures of the graphical method of moments were named after their inventors Wilhelm Ritter and Karl Culmann. Although Ritter's method has an analytical counterpart, Culmann's method is strictly graphical. Graphical methods were only occasionally used for bridge trusses, because of the variable loading on bridge structures; analysis of trusses by William Merrill's method (1870) and the later efforts of Mansfield Merriman and Henry Sylvester Jacoby (1894) are exceptions. Bridge designers generally preferred the analytical methods discussed in Chapter 7.

Graphical Method of Joints

This method, also known as Maxwell's method or the Cremona-Maxwell method, consists of constructing force diagrams for the bar forces at each joint in sequence. Because a bar connects to two joints, once the force in one end of the bar is known, the force in the other end is also known and can be used to solve the next joint.

The great advantage of the graphical method of joints is that all of the joint force diagrams can be superimposed on the same diagram. This requires the use of a system of notation developed by Robert Henry Bow (1873, p. 51), in which the space between each of the loads and reactions is given a letter of the alphabet, and each panel of the truss is given a number. Each force can be represented by a pair of alphanumerics. The first load from the left-hand side of the truss in Figure 12-1 is *a-b*, the force in the vertical at the center of the truss is 5-6, and the force in the top chord just to the left of the gable is *c*-5. The first set of forces drawn on the diagram are the loads and reactions, starting from *a-b* and closing with *g-a*. When all of these forces are vertical, this is a vertical line, known as the "load line" (shown in Figure 12-2).

When the load line is drawn, any joint with two unknown quantities can be solved. For the truss of Figure 12-1, the joint at the left or the right support has to be solved first. To solve this joint, take the forces at the joint in a clockwise direction, beginning with known forces and continuing to unknown forces. At the left support, the 2,500-lbs reaction, *g-a*, is known, and *a*-1 and 1-*g* are unknown. The point 1 must be at the intersection of the directions of line *a*-1, at a slope of 6 in 12 from *a*, and line 1-*g*, in a horizontal direction from *g*.

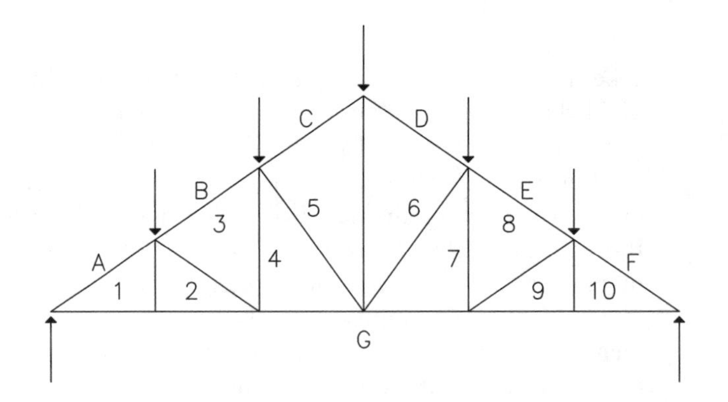

Figure 12-1. Inclined chord Howe truss to be analyzed by graphic methods.

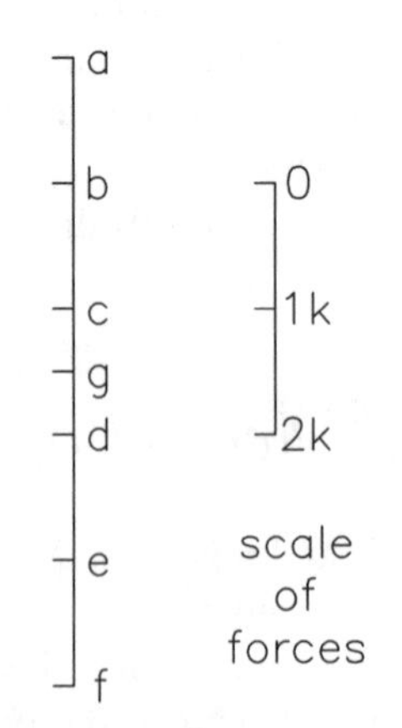

Figure 12-2. Load line for truss analysis.

The force in *a*-1 can be read off as 5,000 horizontal, 2,500 vertical, or measured as 5,590 lbs. The force in 1-*g* is 5,000 lbs (see Figure 12-3).

Joint *g*-1-2 is next, with 2 being located on a vertical line from 1 and a horizontal line from *g*. Because 1 already lies on a horizontal from *g*, this can only be constructed if 1 and 2 coincide, making 1-2 a 0-force member. With 1 and 2 known, the next joint to solve is 2-1-*a*-*b*-3, with unknowns *b*-3 and 3-2. Unknown *b*-3 has a slope of 6 in 12, going up from left to right, and 3-2 has a slope of 6 in 12, going up from right to left, enabling the construction of point 3. The force in *b-3* is 4,000 horizontal and 2,000 vertical, and the force in 3-2 is 1,000 horizontal and 500 vertical (see Figure 12-4).

Continuing in like manner through the center of the truss, the diagram of the forces to the left of the gable, shown in Figure 12-5, can be constructed. Symmetry allows quick completion of the diagram for the truss, as shown in Figure 12-6.

Among many others, this method is described thoroughly by Luigi Cremona (1890), Swain (1896), Charles Ezra Greene (1877), Milo Ketchum (1903), Frank Kidder (1886), Merriman and Jacoby (1894), and nearly every other contemporary writer on graphic statics.

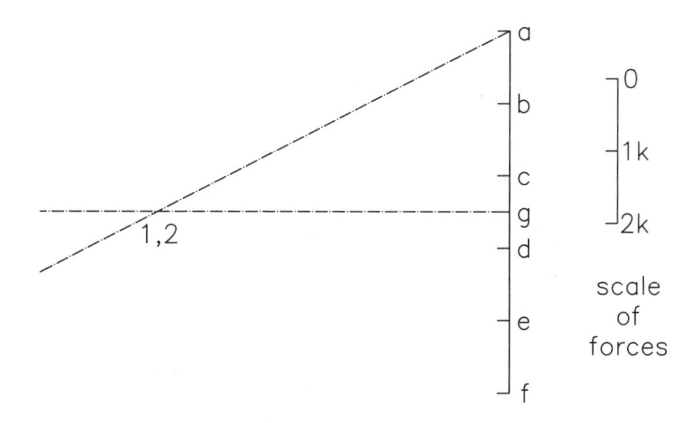

Figure 12-3. Analysis of support joint by graphic method.

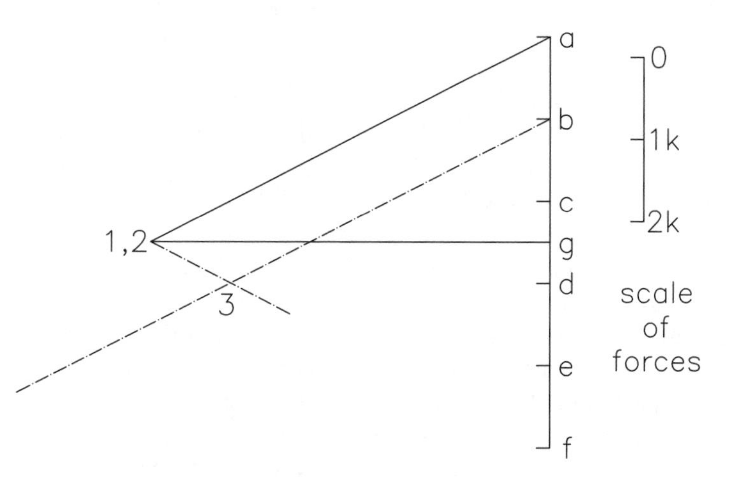

Figure 12-4. Analysis of second panel by graphic method.

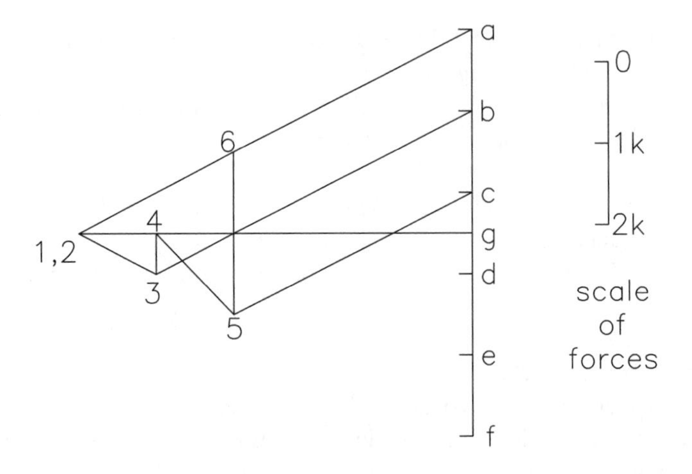

Figure 12-5. Analysis of half-truss by graphic method.

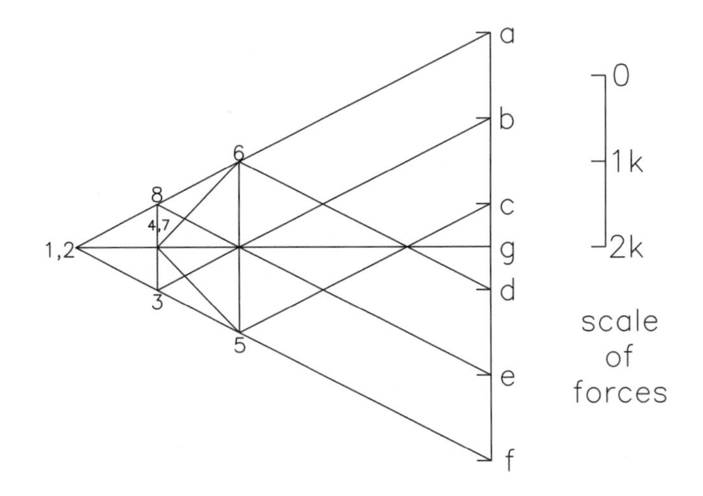

Figure 12-6. Analysis of complete truss by graphic method.

The method was preferred for the analysis of building trusses through at least the 1930s when more widespread calculating machinery made the analytical method of joints preferable. Examples of the use of this method include bachelor of engineering theses from Cornell and Lehigh Universities (Shumway 1904, Dunnells 1897).

Kidder (1886, p. 436) presents extensive graphic analyses of commonly used roof truss forms (Figure 12-7). His analysis follows exactly the method as presented. He applies this method to various roof trusses from simple trusses to trusses with curved top and bottom chords. The design of such trusses in wood and iron is accomplished according to the rules previously described for the tensile and compressive resistance of these materials. Merriman and Jacoby (1894, part 2, p. 49) extend the method to trusses loaded by wind pressure only. In the method they present, a degree of statical indeterminacy at the supports is allowed and overcome by assuming the direction of the reaction at one of the supports is in the direction of the normal wind pressure on the windward side of the roof (Figure 12-8).

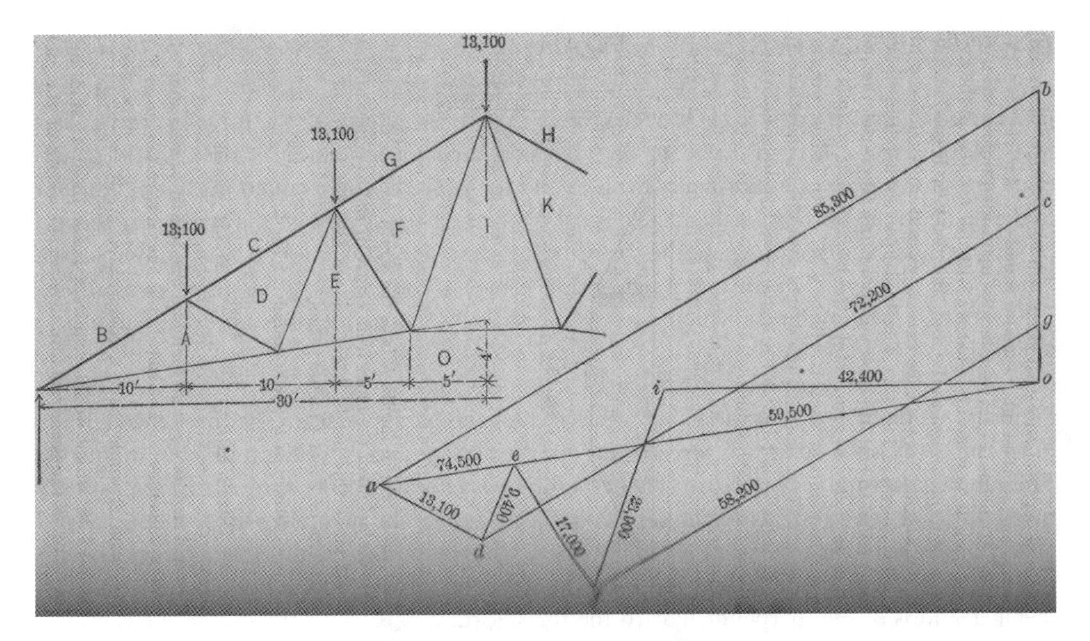

Figure 12-7. Analysis of a roof truss for gravity loading.
Source: Kidder (1886).

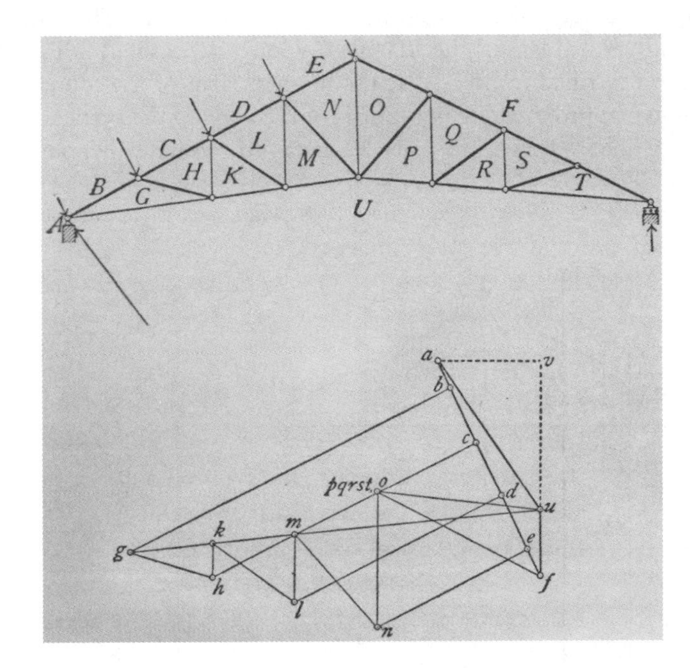

Figure 12-8. Merriman and Jacoby approach to wind load design of a roof truss. Loading and truss configuration are shown in the upper figure, with assumed direction of windward support reaction. Computation of bar forces in the truss by Cremona-Maxwell method is shown in the lower figure.
Source: Merriman and Jacoby (1894).

Graphical Method of Moments

It is not necessary to analyze each of the joints in turn to determine the force in a given bar in a complex truss. It is sufficient to cut a section through a portion of the truss in which there are three or fewer unknown bar forces and to apply the three equations of equilibrium in the plane to the portion of the truss cut off by the section. Although the analytical version of this method is now known as the "method of sections," this method was known as the "method of moments" in the late nineteenth century. In the following material, two graphical versions of the method of moments are presented, named for their authors, Ritter and Culmann.

Ritter's method consists in cutting a section through three bars of a truss, then taking moments graphically about the point where two of the bars intersect. The force through the third bar is the only unknown force acting on the system and can be found from moments. The moment acting on the truss can be computed graphically, as shown in Figures 12-9 through 12-11 for a four-panel Fink, or Polonceau, truss. The moment arm of the top chord at section *b* about the intersection of the bottom chord and the web member coincides with the first web member. This line joins the intersection of the bottom chord and the web member and is drawn perpendicular to the top chord.

In Figure 12-9, Ritter's method is used to find force in the top chord at section *b*. First the force and funicular polygons for the truss and its loading are drawn. By Culmann's theorem (see Chapter 11), the moment at section *b* equals the horizontal distance from pole to load line in the force polygon times vertical distance between the two strings on the funicular polygon on a vertical line from the center of moments. These quantities are laid out as the base and vertical legs of the triangle (hypotenuse dashed) in Figure 12-10. Using the transformation described in Box 12-1, the moment arm of the top chord about center of moments is shown approximately two-thirds of the way up the vertical leg of the triangle in Figure 12-10. A line is drawn in Figure 12-10 as to the inner hypotenuse beginning at

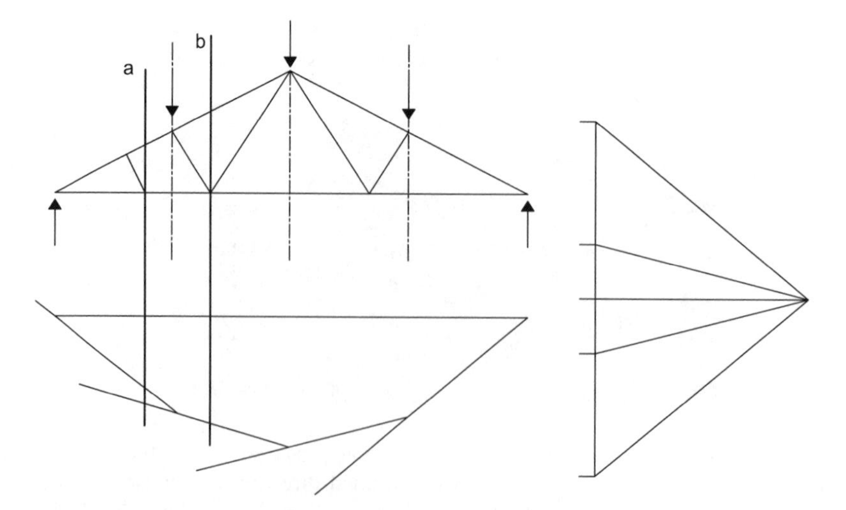

Figure 12-9. Force and funicular polygons for a four-panel Fink truss.

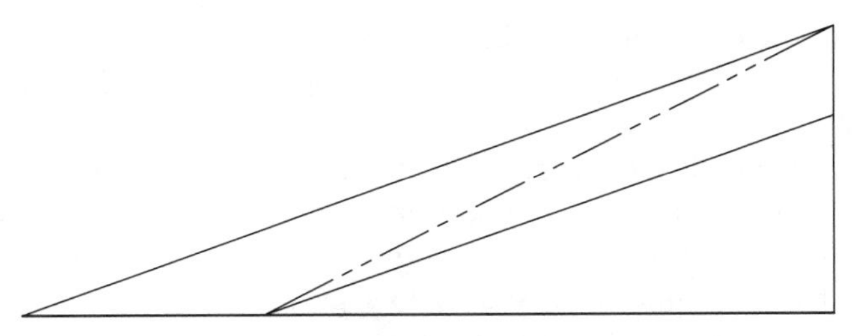

Figure 12-10. Graphical determination of top chord force adjacent to gable, four-panel Fink truss.

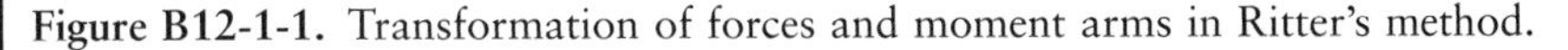

Box 12-1

To explain the application of Ritter's method, some preliminaries are necessary (Culmann 1875). Moments can be represented graphically on the funicular diagram by drawing a scaled force perpendicular to a given moment arm. An example shown in Figure B12-1-1 shows force OF and moment arm Or. Then, the area of the triangle drawn on these two lines represents (half) the moment of the force. To find the force associated with a different moment arm, say Or' in Figure B12-1-2, draw a line through r parallel to Fr'. The intersection of this line with OF, extended, gives OF' such that $OF' \times r' = OF \times r$.

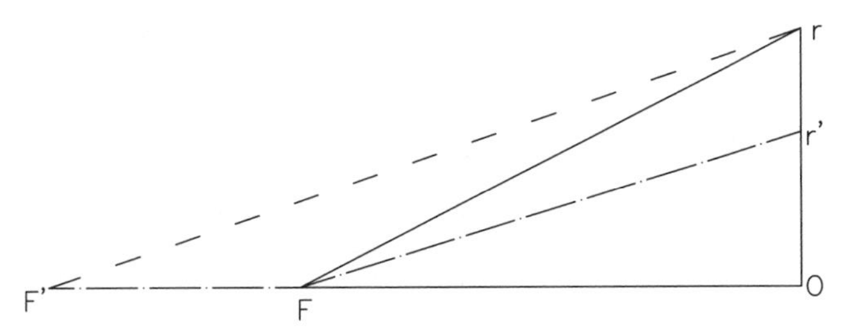

Figure B12-1-1. Transformation of forces and moment arms in Ritter's method.

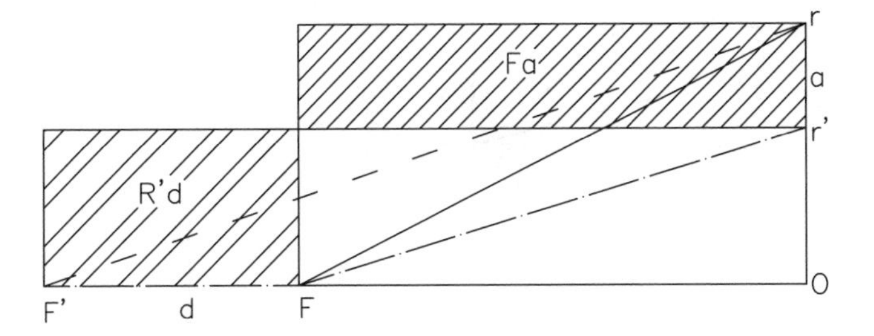

Figure B12-1-2. Proof by areas of transformation of forces and moment arms in Ritter's method.

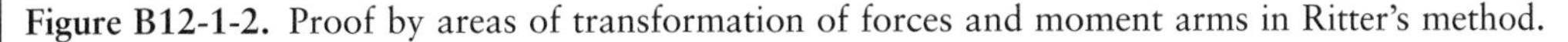

The geometrical proof of this transformation is given in Figure B12-1-2. The two hatched areas are equal. Let a = the difference between r' and r and d = the difference between F' and F. The hatched area to the left = $r'd$, and the hatched area above = Fa. However $a/d = r'/F$.

Substituting $r' = aF/d$, the areas are equal. By subtracting one area from RF and adding it to $r'F$ gives $r'F'$, we can conclude that $rF = r'F'$.

the moment arm of the top chord. The intersection of the outer hypotenuse (parallel to the inner) with the horizontal defines the force in the top chord. The bottom chord force can be found similarly.

By Culmann's method, the top chord force at a section is found by decomposing the resultant force (read from the force diagram) into a force along the top chord and a force passing through the point of intersection the other two members cut in the section (bottom chord and web). For the same truss shown in Figure 12-9, Figure 12-11 illustrates this process. On the force polygon, the resultant force at the section is found on the load line. The line of action of the resultant force is represented on the funicular polygon as passing through the point of intersection of the two strings representing components of the resultant force, well to the left of the support point. Following this step, the resultant force can be decomposed into a force x parallel to the top chord (to the left of the load line) and a force y passing through the intersection of the two other bars cut by the section. These directions are transferred to the force polygon, determining the forces in the top chord. As the force y is the resultant of the forces in the other two bars, their directions can be represented as components of y. This decomposition is also shown on the force diagram in Figure 12-11.

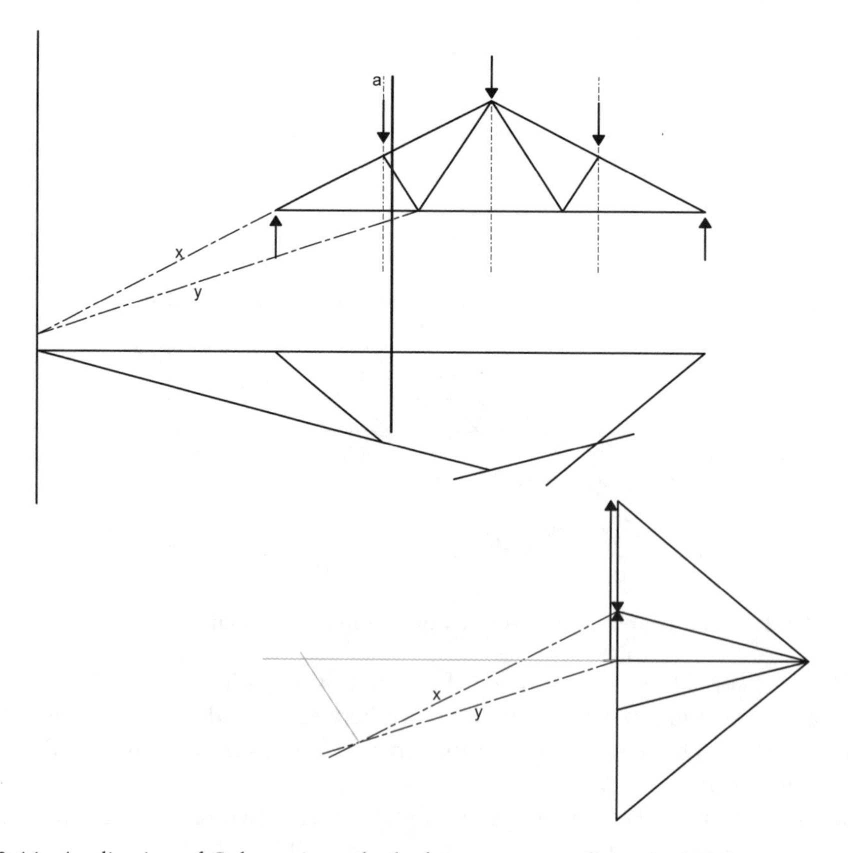

Figure 12-11. Application of Culmann's method of moments to a four-panel Fink truss.

Semigraphical Methods for Analysis of Bridge Trusses

Hermann Haupt (1858), in investigating simple wooden bridge trusses in a Howe configuration, applied a simple semigraphical method to the determination of the forces in the chords under single-panel loading, which is extended by Merrill (1870), whose method is described following. Merrill's method consists in finding the bar forces due to an individual panel load and summing all of the forces in each bar due to all of the panel loads present on the bridge. The forces due to a panel load are found by a simple graphic construction of the funicular diagram of the support reactions and the panel load and subsequently decomposing the two strings into components along all of the bars at the panel joint. This method is best illustrated by a simple example. For the six-panel Pratt truss shown in the upper diagram in Figure 12-12, the direction of the reactions due to a force applied at the second panel point can be found by drawing lines to the points of support. By transferring the direction of these two resultant forces by the parallelogram rule, they are used to represent, by means of the upper line segments, the two components of the unit load applied at the panel point. Each of these line segments can be further decomposed into forces in the directions of the bottom chords, tie, and vertical strut, as illustrated in Figure 12-12. The figure taken from Merrill shows the further process of constructing similar diagrams at each panel point and determining the force in any member of the truss due to a unit load at a specific panel point. The actual determination of bar forces under a given train of loading consists of superimposing

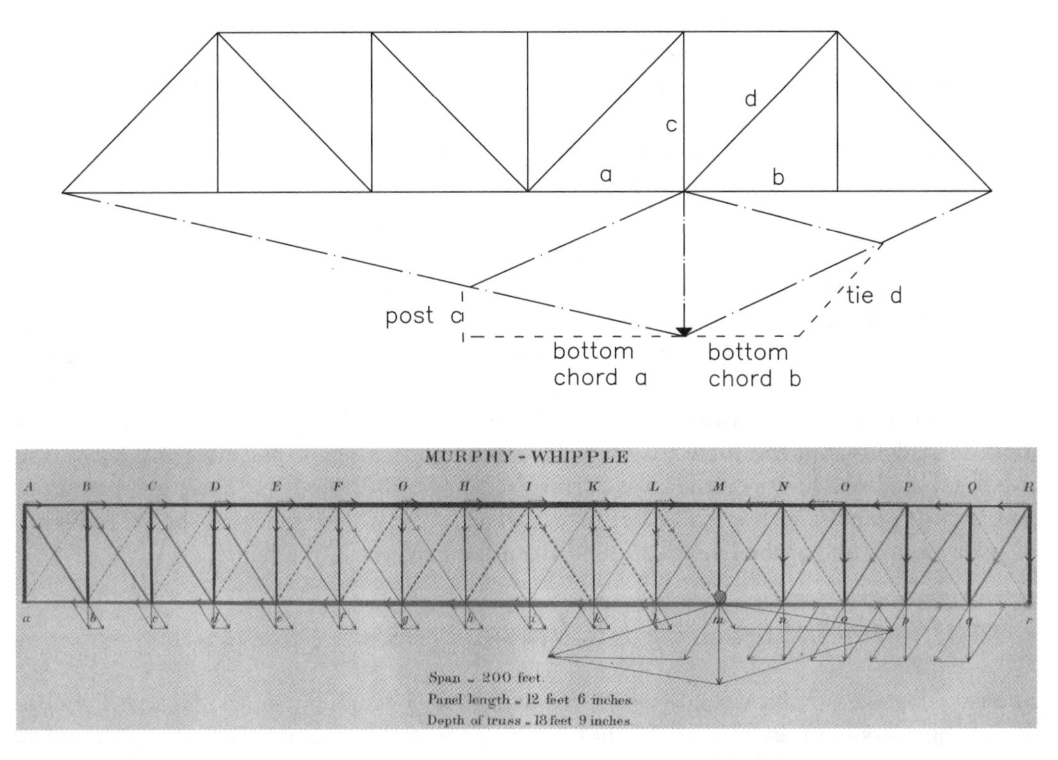

Figure 12-12. Example of semigraphical analysis of a Pratt truss by Merrill.
Source: Merrill (1870).

TABLE 12-1. *Bar Force Multipliers for Web Bars in an Eight-Panel Warren Truss*

Bar (numbered from left support)	Maximum tensile live load multiplier	Maximum compressive live load multiplier
1	0	35
2	35	0
3	1	21
4	21	1
5	3	15
6	15	3
7	10	6
8	6	10

the forces due to all of the panel point loadings that result from the weight of the locomotive and train.

Merrill's semigraphical discussion of detailed truss designs is used similarly by William Eads (1868) in a discussion of the action of a truss, meant to show that Eads's arch design for the St. Louis Bridge was more economical.

Merriman and Jacoby (1894) present a different semigraphical method of bridge truss analysis. Dead load stresses are found in the same manner as stresses on bridge trusses, by drawing a Cremona-Maxwell diagram for the truss. Merriman and Jacoby further point out that maximum chord stresses result from all panels loaded condition and also can be calculated by the previous graphic method. However, because of the variable nature of live load stresses in truss members, a modified procedure needs to be followed for the live load stresses in the web members. To determine maximum web stresses due to live load, Merriman and Jacoby note that it is only necessary to draw the web stresses for one load position and to recognize that for other load positions the web stresses increase by an integral factor.

For instance, given an eight-panel Warren truss, such as that shown in Figure 12-13, a unit load positioned as shown can be found graphically to result (in the dashed web member, three panels from the left support) in the force shown dashed on the force diagram. For a load positioned at the next position to the left, the force in the web member under consideration will double. For a force positioned at each of the panel points to the right of the web member under consideration, the total force will be 5 + 4 + 3 + 2 + 1, or 15 times the live load force in the force diagram. It is a simple matter to assemble a table of the maximum live load forces on the basis of the force diagram. Table 12-1 shows the maximum tensile and compressive live load force on each bar of the web of the truss shown in Figure 12-13 as a multiple of the live load found from the diagram in the figure.

Graphical Analysis of Building Trusses by Swain

In describing the graphical analysis of various forms of building trusses, George Fillmore Swain considers both the method of joints and the method of sections (1896). He describes the process of computing a truss for design, requiring the completion of four truss diagrams, for dead load, snow on left, wind on left, and wind on right, combining the load effects as

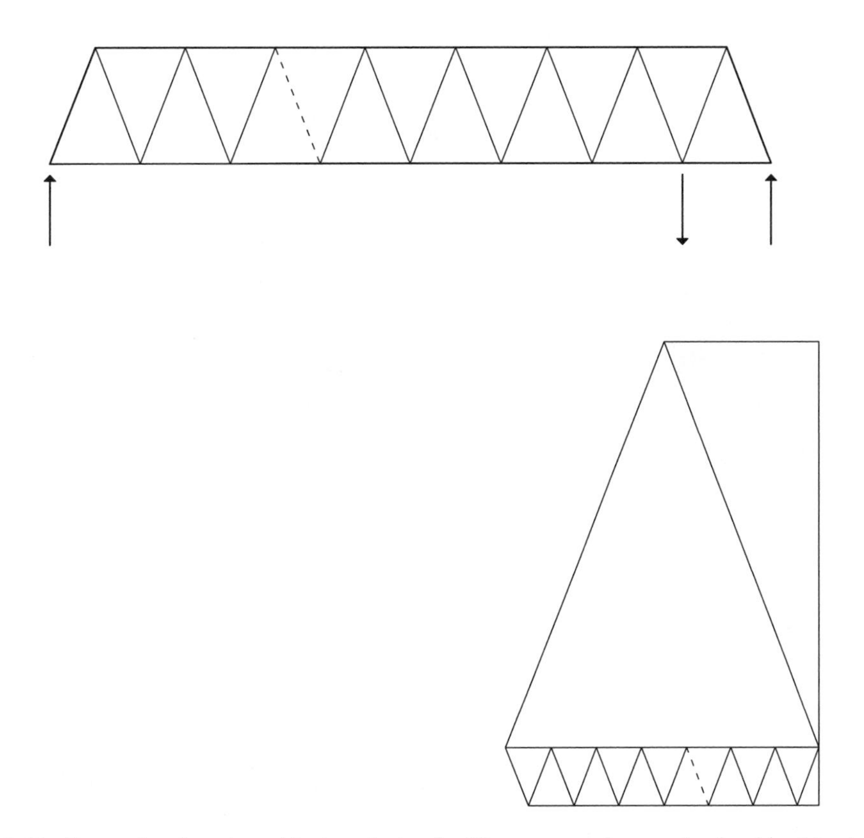

Figure 12-13. Example of semigraphical analysis of a Warren truss for moving load by Merriman and Jacoby.
Source: Merriman and Jacoby (1894).

necessary. He shows these diagrams for a gabled Howe truss; a gabled Warren truss with a cambered bottom chord; an eight-panel Fink, or Polonceau, truss; and a pair of fan trusses: six-panel and 12-panel.

Both the Fink truss and the fan trusses present a problem in analysis, which Swain terms "ambiguous cases." Although this discussion is available in Swain's 1896 course notes, it is easier to find in his 1927 textbook. A different procedure to solve the same problem appears in Merriman and Jacoby (1894), part II, page 61. In general, it is only possible to solve for two unknown forces at a joint by Maxwell's method. Most trusses can be solved directly in this fashion. A few types of statically determinate trusses have joints with more than two unknown forces when proceeding joint by joint. In the eight-panel Fink truss shown in Figure 12-14, it is possible to solve joints *a*-1-*i*, *a*-*b*-2-1, and 1-2-3-*i* directly. However, the following joint *b*-*c*-5-4-3-2 has three bars with unknown forces.

Swain's procedure illustrated using the left half of the truss in Figure 12-14. Construct 1, 2, 3 normally, then assume a location for 5 on a line following the roof pitch to C (shown as *5* on the partial force diagram at the lower left). Find 6 on the basis of *5* (follow slope of 5-6 to diagonal from *D*) and 7 on basis of 6. Point 4 must be on diagonals from 3 and 7 (based on slope of 3-4 and 4-7). Point 4 must also be on a horizontal from 5, and 7 must remain on the horizontal from *I*, and 6 must remain on the sloped line from *D*. These

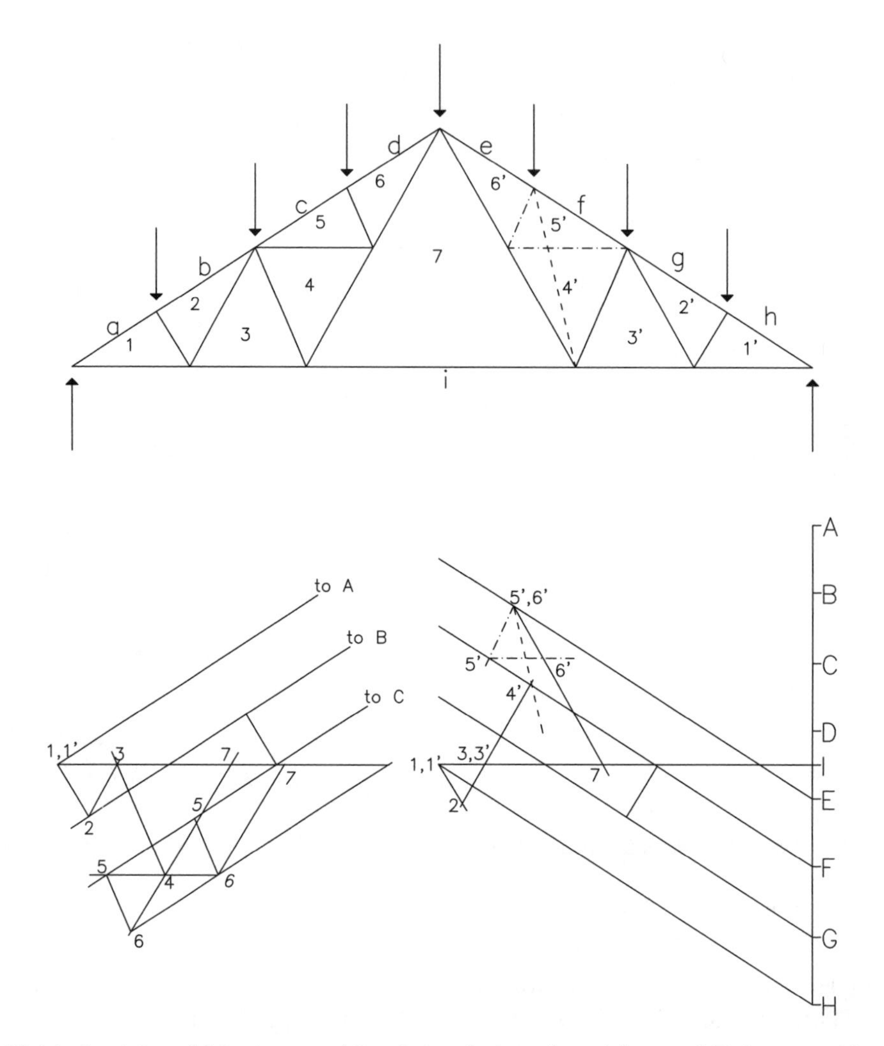

Figure 12-14. Swain's and Merriman and Jacoby's solution of an eight-panel Fink truss with ambiguous members.

conditions can only be satisfied by shifting the triangle 5-6-7 to the left, as shown. This can then be redrawn on the truss diagram.

The second procedure for the solution of this system is given by Merriman and Jacoby (1894). Consider an auxiliary member (drawn dashed) to temporarily replace the actual diagonal 2′ to 3′. The truss can be solved based on this member (shown on the lower half of the truss diagram), suppressing panel 5′ temporarily. This procedure correctly locates point 7. Points 6′ and 5′ can be correctly located on the line 5′, 7, and the remainder of the points for the actual truss follow immediately from this. The completed diagram for this truss is shown on the right half of the truss in Figure 12-14. A complete solution of the eight-panel Fink truss is shown in Figure 12-15.

Swain (1896) also describes the graphical method of moments for a similar truss. In the upper diagram in Figure 12-16, Swain determines the forces by the calculation of

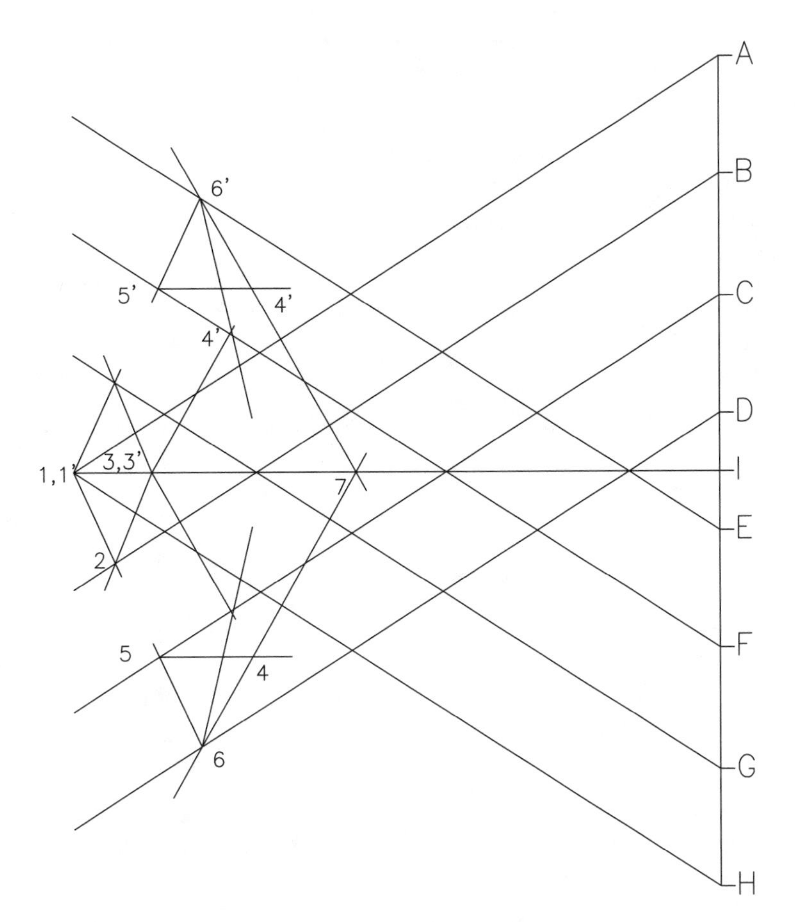

Figure 12-15. Complete graphical solution of eight-panel Fink truss.

moments and the transformation of the moment arm, using Culmann's theorem, through an application of Ritter's method. For instance, at a section through the first panel, he chooses the moment arm l_6 to determine the force in the bottom chord. The global moment on the truss at that location is found by Culmann's theorem to be $H \times a_6b_6$ on the funicular polygon. To find this moment divided by l_6, he lays off l_6 perpendicular to a_6b_6 at b_6 He then draws a similar triangle to $a_6b_6l_6$ with base of length H at c_6d_6. Then, by similar triangles $H \times a_6b_6 = l_6 \times a_6d_6$, so that a_6d_6, measured on the scale of forces, is equal to the force in the bottom chord at section. The diagram includes a similar graphical analysis of all the bars in the truss.

In the lower diagram in Figure 12-16, Swain undertakes an analysis of the same truss under the same loading conditions by Culmann's method. The force in the bottom chord bar 5-6 is found by locating the resultant of the forces acting on the left side of the section shown lightly on the drawing of the truss; this is done by extending strings 0-1 and the closing line to their intersection. The resultant, passing through this point, is decomposed into *s-t*, passing through the point of intersection of the top chord and the web member, and *s-t'*, parallel to the bottom chord. When transferred to the force polygons, these

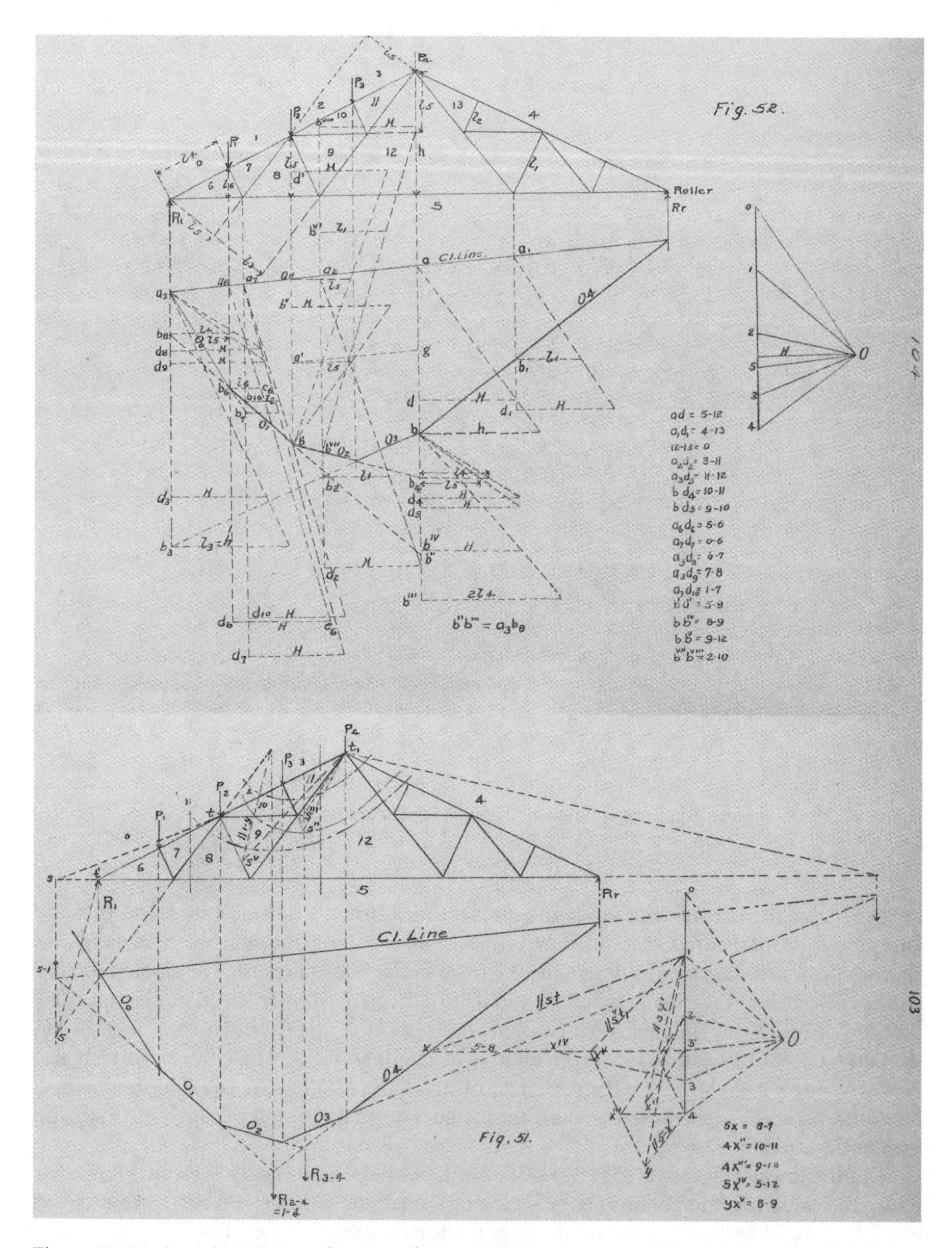

Figure 12-16. Swain's (1896) application of Ritter's (upper diagram) and Culmann's (lower diagram) methods to the analysis of an eight-panel Fink truss.

Source: Reproduced by permission from The Huntington Library, San Marino, CA. Call No. Rare Books 624550.

directions yield the force in the bottom chord. Similar operations are performed at other sections throughout the truss.

Use of Graphical Methods of Truss Analysis

In the late 1800s, many graphic procedures were available for the determination of forces and deflections in trussed structures. These were widely used, depending on the ease of use and the perceived usefulness of the information extracted from the method. The construction of force diagrams for building trusses was a method widely used, because it was considered significantly simpler than its analytical counterpart. This method was used generally for the analysis of building trusses. Although the graphical method of moments was well known and taught in many engineering schools, these methods do not appear to have been so widely applied. An examination of theses from engineering schools on the East Coast from the late 1890s shows a widespread application of graphic methods to the solution of building trusses. In the design of mill buildings, it was necessary to apply these methods to the analysis of braced bents that incorporated a roof truss. This topic will be described in more detail in Chapter 15.

References Cited

Bow, R. H. (1873). *Economics of construction in relation to framed structures.* E. and F.N. Spon, London.

Cremona, L. (1890). *Graphical statics: Two treatises on the graphical calculus and reciprocal figures in graphic statics.* Trans. T. H. Beare, Clarendon Press, Oxford.

Culmann, K. (1875). *Die graphische statik,* 2nd Ed. von Meyer and Zeller, Zurich (in German).

Dunnells, C. G. (1897). *Design for a steel mill building.* Bachelor's thesis (manuscript) in civil engineering, Lehigh University, Bethlehem, PA.

Eads, W. B. (1868). *Report of the engineer-in-chief of the Illinois and St. Louis Bridge Company.* Missouri Democrat Book and Job Printing House, St. Louis, MO.

Greene, C. E. (1877). *Graphical method for the analysis of bridge trusses.* John Wiley and Sons, New York.

Haupt, H. (1858). *General theory of bridge construction.* D. Appleton, New York.

Ketchum, M. (1903). *Design of steel mill buildings.* Engineering News Publishing, New York.

Kidder, F. (1886). *The architects' and engineers' pocket-book,* 3rd Ed. John Wiley and Sons, New York.

Merrill, W. (1870). *Iron truss bridges for railroads.* Van Nostrand, New York.

Merriman, M., and Jacoby, H. S. (1894). *A treatise on roofs and bridges, part 2*, 2nd Ed. John Wiley and Sons, New York.

Sahag, L. M. (1958). *Applied graphic statics.* Edwards Brothers, Ann Arbor, MI.

Shumway, A. K. (1904). *Wind stresses in a steel frame mill building.* Bachelor's thesis in civil engineering, Cornell University, Ithaca, NY.

Swain, G. F. (1896). *Notes on the theory of structures*, mimeographed course notes from lectures presented at the Massachusetts Institute of Technology, Department of Civil Engineering, 2nd Ed. Published in book form in *Structural engineering. Stresses, graphical statics, and masonry.* McGraw-Hill, New York, 1927.

Turner, W. W. (1966). *Elements of graphic statics.* Ronald Press, New York.

13

Graphical Analysis of Arches

Graphic analysis of arches consists of locating the internal compressive resultant within the arch under analysis or design. This line is alternately known as the thrust line or the line of resistance. The funicular polygon construction, which is described in Chapter 11, is particularly useful for constructing the thrust line or line of resistance of an arch. The thrust line generally is defined (Heyman 1982) as a line composed of points in an arch where the external forces (loads and reactions) can be resisted without moments. This idea is illustrated for a section of an arch in Figure 13-1. If the reactions are known, a funicular polygon automatically determines the thrust line of an arch. From the construction of the thrust line of the arch, it is possible to determine the magnitude of the internal axial force from inspection of the force polygon. It is more convenient to calculate the bending moment directly as a product of the resultant force at a point and the deviation of the thrust line from the centerline axis of the arch (see Chapter 11). An alternative, for an arch subjected to vertical loads only, is to formulate Culmann's theorem (Chapter 11) based on the horizontal component of the reaction, equivalent to the pole distance in a beam problem, and the bending moment at any point in the arch becomes the pole distance multiplied by the vertical distance between the funicular polygon and the centerline of the arch (Sondericker 1903, p. 74).

Moreover, by confining the thrust line to a predetermined portion of the arch ring, the stress in the arch can be mitigated. Although the middle one-third of the arch ring was often considered the proper limit on the thrust line location for an arch, less restrictive limits were observed by some authors, according to the discussion in the final section of this chapter. Because the determination of reactions is

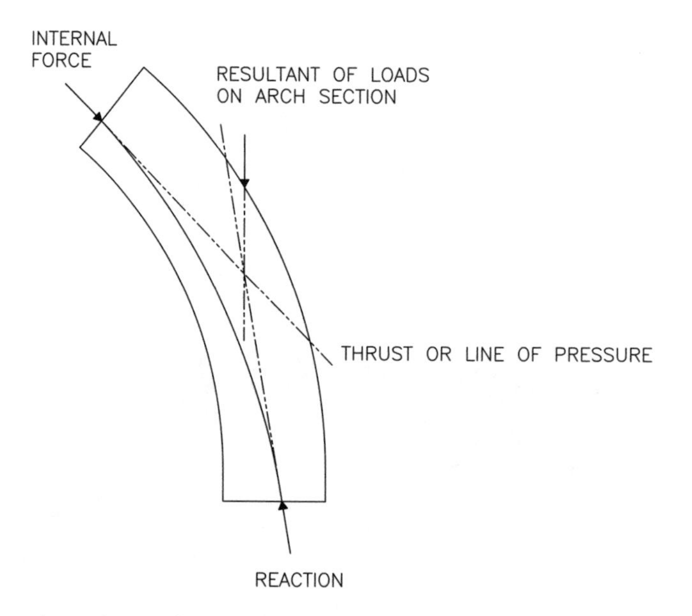

Figure 13-1. Thrust line of an arch at a cross section.

approximate and is based on balancing the eccentricity of the thrust in some fashion, having the thrust line available allows the analyst to achieve this balance visually, resulting in a close approximation to the supposed actual thrust line.

Some methods for the determination of an arch thrust line are reviewed in this chapter; however, trial and error remains the most generally applicable method. In most cases, the required analysis is for a symmetrical arch, with the result that only half the force polygon needs to be used and that the string at mid-span, representing the crown thrust of the arch, is horizontal. Figure 13-2 shows the general procedure. Divide the arch into sections, then determine the weight and locate the centroid of each section. The weights of the sections are put in order into a force diagram. Choose a pole location: for a symmetric arch, the pole can be such that the crown thrust is horizontal, and rays are drawn from the pole to the intersection points of the loads on the load lines. Transfer the direction of these rays to the funicular diagram in such a way that the two rays that are components of a given load intersect on the line of action of that load. The location of the resultant of the forces can be found on the funicular diagram by joining the two external strings, as shown in Figure 13-2.

Masonry arches lend themselves to graphical analysis for three important reasons. First, a fixed arch is, in general, three degrees statically indeterminate and, on these grounds alone, particularly difficult to solve analytically. The approximate analytical methods of dealing with static indeterminacy in arches given in Chapter 6 demonstrate how tedious this type of analysis can be. Second, the curve of the arch complicates transforming vertical loads into the coordinate system of the arch. As a result, even statically determinate arches have relatively complex formulas to determine the forces in the arch. Third, for masonry arches, the lack of tensile resistance dictates distinct bounds on the solution that make approximate solutions credible. If the masonry is assumed to have no tensile resistance, then the line of

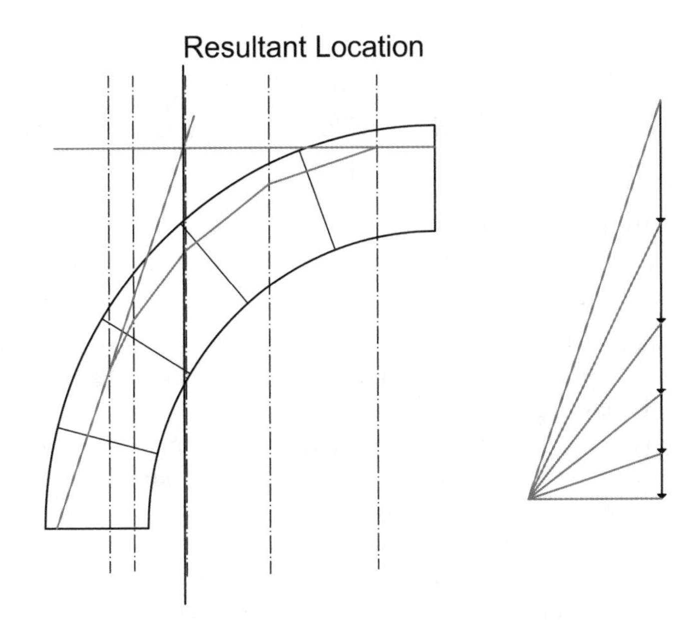

Figure 13-2. Simple analysis of a symmetrical arch.

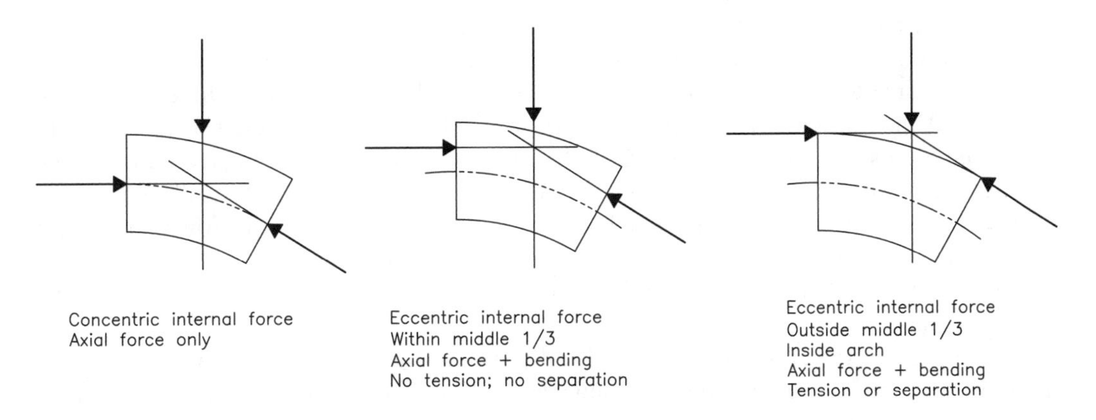

Figure 13-3. Various conditions of pressure within a segment of an arch.

pressure in the arch is required to be within the arch, whereas if it is considered undesirable to generate tensile stresses, the line of pressure must be within the middle third of the arch (see, for instance, Baker 1907, p. 449). This principle is illustrated in Figure 13-3. The middle third condition may be considered a prerequisite for stability of the arch, or alternatively, it may be considered a desirable state for serviceability: many nineteenth-century authors do question the merit of this criterion as a stability condition. In fact, no arch to this author's knowledge is able to maintain the "middle one-third rule" under all expected loading conditions. In the words of Ira Osborn Baker (1907, p. 451), rebutting William John Macquorn Rankine's assertion of the middle third rule, "A reasonable theory of the arch will not make a structure appear instable which shows every evidence of security." George F. Swain (1896, p. 123) gives the following conditions of safety of the arch:

- The true line of resistance must be within the arch ring or with the middle third if there is to be no tension.
- The true pressure on any joint must not make the normal to that joint a greater angle than the angle of repose.
- The true maximum intensity of pressure at any joint must not exceed the allowable stress.

However, in spite of the first statement allowing the thrust line to stray from the middle third, Swain applies the middle third rule to the determination of the suitability of an arch in all his subsequent constructions.

The approximate analysis of the arch reduces to finding a suitable line of pressure within the arch for the given loads, an exercise that can be completed graphically by the construction of a funicular polygon appropriate for the given loads imposed on the structure. This construction can be done for dead or construction loads separately from dead plus live load, or for various configurations and locations of live load. The basis of the procedure is to divide the arch and its loading into segments, to use the series of these load segments for the construction of a force polygon, and to use the force polygon to complete a funicular polygon. The correct funicular polygon can be chosen by trial and error or by completing a single funicular polygon using a trial pole location and using the properties of the funicular polygon to determine the required funicular polygon.

Several theories, previously described in Chapter 6, were advanced for the determination of the correct line of resistance, or thrust line, for the arch. These included the determination of the minimum least squares error of the thrust line from the centerline of the arch, a laborious mathematical procedure that, if it was used, was probably applied intuitively by minimizing the deviation about the centerline of the arch. The least absolute pressure was also used as a criterion, but Baker calls this "a meta-physical principle." Most common (because easiest to apply) was to use the least permissible crown thrust. Although this has been considered as a principle to be applied analytically, it is also amenable to graphical analysis.

The basis of the application of graphical statics to the determination of stability in a masonry arch is the construction of a funicular polygon for the given loads (see Chapter 11) that lies within the arch. Although we will consider primarily symmetric loading cases, the method is equally applicable to nonsymmetric loading. The arch is subdivided into segments: if the arch and the loading are symmetric, it suffices to consider half the arch, noting that the line of pressure will be horizontal at the axis of symmetry. If only vertical loads are considered (no horizontal earth pressure), a vertical load line is drawn to a force scale. A pole is selected and the polar rays are drawn to the points of intersection of the loads on the load line. This initial pole may be selected on a horizontal line from the point of symmetry of the load line, or it may be selected arbitrarily. By constructing the polar rays on the funicular diagram as the components of the load segments, it is possible to determine a funicular polygon appropriate for the loads on the arch. As the objective of the construction is to draw a funicular polygon within the arch (or within the middle third of the arch), the pole location must be moved, and a correct funicular polygon drawn, either by trial and error or by a more efficient procedure, such as Méry's method.

In Méry's method, the initial choice of a pole is arbitrary, resulting in an arbitrary funicular polygon. The intersection of the two strings representing the crown thrust and the support reaction gives a point on the line of action of the resultant of all the loading. Then a horizontal line can be drawn from the crown, representing the crown thrust, and an inclined line can be drawn from the abutment intersecting the line of action of the resultant at the same point as the crown thrust. The construction of the funicular polygon can be completed, either by locating the correct pole location on the basis of these two external strings or by constructing the remaining strings directly by incidences. The basis of the latter method is described later in this chapter. This procedure is illustrated in Box 13-1 and the accompanying figures are taken from Frank Kidder's *Architects' and Builders' Pocket-Book*. (1886).

Hermann Haupt (1856, p. 137) citing Thomas Tredgold, shows a particularly simple method of constructing the thrust line where the two resultants are located, and on the simple assumption of a parabolic thrust line, it is possible to construct by tangents. Haupt's point of view on where the thrust line should be to ensure the stability of the arch is "the load upon the different parts of the arch and the curve of its intrados must bear such a relation to each other that the line of pressure will never fall outside the limits of any joint, but will approach as nearly to the center of the joints as possible."

In Haupt's construction, having found or estimated the position of the resultant, the support reaction, and the crown thrust, the two segments of the crown thrust and support reaction are divided into a number of segments, equal to the number of segments between the loads (in the case shown in Figure 13-4, three). The tangents to a parabola may be constructed by connecting the point farthest from the resultant on one line segment with the point closest to the resultant on the other (shown in Figure 13-4) and proceeding

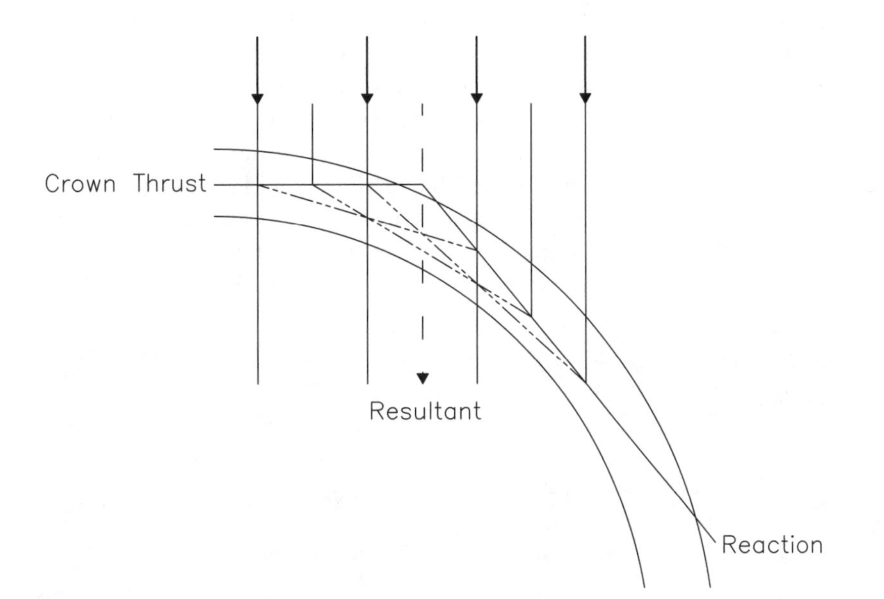

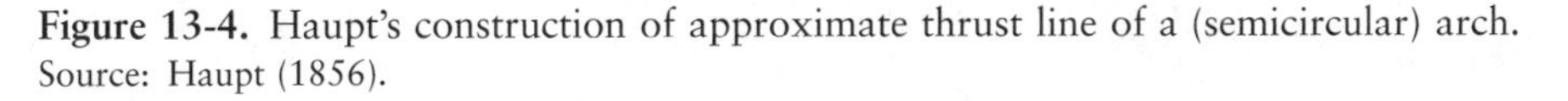
Figure 13-4. Haupt's construction of approximate thrust line of a (semicircular) arch.
Source: Haupt (1856).

Box 13-1. Construction of Line of Pressure

This example, taken from Kidder (1886), presents the solution of the line of pressures for a semicircular arch symmetrically loaded only with self-weight in two steps: finding the joint of rupture and solving for the arch between joints of rupture. In the left-hand portion of Figure B13-1-1, the half arch is divided into 10 equal segments, and the load on each segment computed and shown on the load line as A-1, 1-2, and so on. An arbritrary initial selection of pole is made (O), and the preliminary funicular polygon is drawn, the catenary line *a, b, ..., n*. By extending *ab* and *kn* to their point of intersection, the line of action of the resultant is found to pass through O on the funicular diagram. Then, drawing a horizontal line from the middle third point at the crown *A* (producing least crown thrust) and locating its intersection with a line through *B*, the exterior middle third point at the abutment, the pole O on the force diagram can be determined. Filling in the rest of the polar rays and transferring to the funicular diagram results in the funicular diagram as shown. It is noted that this diagram passes out of the middle third of the arch, with the error being greatest at the seventh joint. Hence, the seventh joint is identified as the joint of rupture, and a modified arch is considered in the figure on the right. Here, the same construction is completed with the abutment moved to the seventh joint, and a funicular polygon is found that is contained within the middle third or the arch. Note that the crown thrust had to increase in the second construction. It is remarkable that the force diagram and both funicular diagrams are combined into a single compact drawing in this construction. Although the procedure for constructing a modified funicular polygon by intersections, without redrawing the force polygon is not used here, this procedure is shown later in the discussion of the analysis of Union Arch. Also noteworthy is Kidder's empirical statement regarding the second construction, "This time we see that the line of resistance lies within the middle third, except just a short distance at the springing, and hence we may consider the arch stable" (Kidder 1886, p. 193). Evidently, it is considered permissible for the thrust line to leave the middle third of the arch past the joint of rupture. The application of this method to building design does not generally require the consideration of nonsymmetric loading. A more complex load case for a symmetrically loaded arch is also considered by Kidder, and the combination of the empirical rules presented in this chapter and this graphic method would equip an architect to design most masonry arches that occur in practice. This method is applicable to any symmetric arch, symmetrically loaded. It considerably simplifies the task of locating a thrust line that passes through a desired point at the abutment and at the crown of the arch.

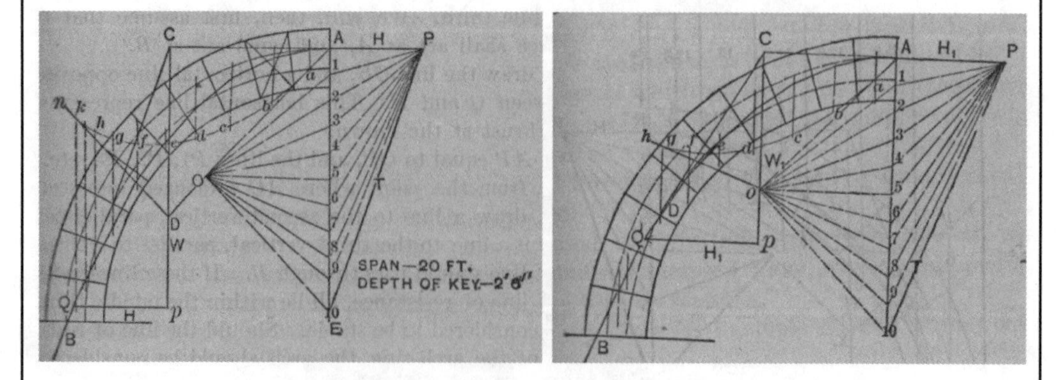

Figure B13-1-1. Kidder's construction of line of resistance for a semicircular arch.
Source: Kidder (1886).

sequentially to connect all of the points. This construction is exact for equal loads and resembles the version of Méry's method described in this chapter in connection with the Union Arch.

When the construction of a line of pressure is known, it is possible to apply some of the methods outlined in Swain or Baker for achieving a closer solution of the correct line of pressure. To review, Baker considered four primary means of determining the actual line of resistance in an arch:

Least Crown Thrust. This is the method employed by Kidder in the aforementioned example. The minimum horizontal thrust consistent with the line of pressure remaining in the middle one-third of the arch ring is the actual thrust line.

Least Pressure. In the least pressure method, a thrust line is drawn for which the maximum pressure in any joint is at a minimum. Generally this is accomplished by keeping the thrust line as close to the center of the arch as possible.

Scheffler's Theory. Scheffler's theory is a semigraphical procedure for the solution of the least crown thrust theory, in which the absolute minimum thrust is determined by summation of moments through the joints from the crown downward, with the center of rotation considered to be the upper middle-third line of the joint.

Winkler's Theorem. According to Winkler's theorem, the correct position of the thrust line is that in which the least squares sum of the deviations from the arch centerline are at a minimum. Although simple enough in statement, the application of this theorem requires drawing a thrust line, measuring the error at each joint, calculating the sum of the squares, and repeating the process until a satisfactory minimum is found.

Swain (1896, p. 127), citing Baker articles 683 and 684 (1907), asserts that the true line of resistance for a vertically loaded arch is the one closest to the centerline. Whether or not he means closest in a least squares sense is unclear, although he speaks of the sum of vertical deviations.

In practice, Scheffler's theory and the least crown thrust theory are closely related. Both criteria result in a thrust line that is as elevated as possible, usually contacting the extrados of the arch at the crown and close to the intrados at the supports. Similarly, Winkler's theorem and the least pressure principle result in similar thrust line configurations that are closer to the center of the arch at the crown and at the supports. The results of the two different sets of criteria are very little different for semicircular arches but begin to deviate for segmental arches of smaller angle of embrace.

Swain (1896, p. 128) asserts what is now known as the lower bound theorem of plasticity (Heyman 1982) as applied to the arch by saying, for instance, that if a line of resistance can be drawn within the arch, the true line of resistance lies within the arch.

> [I]t follows, therefore, that if a line of resistance can be drawn in the arch ring, the true line of resistance lies within the arch ring: or if a line of resistance can be drawn within the middle third, the true line of resistance will lie within the middle third ... this is only a special case of a general principle, first demonstrated in 1879 by the Italian engineer Castigliano, that in any statically undetermined structure, the condition of stress will always be that corresponding to the minimum work.

For instance, for the segmental arch shown in Figure 13-5, the least crown thrust condition is approximately a parabolic line of pressure passing through the extrados at the crown and

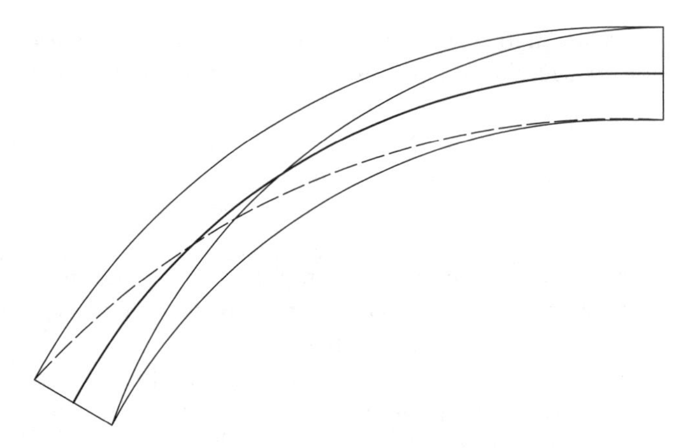

Figure 13-5. Positions of least crown thrust, greatest crown thrust, and least pressure thrust lines.

through the intrados at the support. The greatest crown thrust (shown dashed), however, is an approximately parabolic line of pressure passing through the intrados at the crown and the extrados at the abutment. However, it is also possible to draw a parabola (shown bold) that is very close to the centerline of the arch along its entire length. This represents the least pressure configuration and moreover has significantly lower least squares error (and is thus a closer approximation by Winkler's hypothesis) than either of the other two thrust lines. Malverd A. Howe (1906, p. 34) adds that the line of pressure may well pass out of the middle third, "as long as p [compressive stress] is so small that there is no danger of the stone being crushed the arch is stable. It is a recognized fact that this condition exists in a large number of arches now standing."

The design of a stone arch bridge, according to Swain (1896, p. 128) is to assume the span, rise, and shape of the intrados; then compute the thickness at the crown by one of the empirical formulas from Chapter 2; assume the thickness at the springing and the shape of the extrados; and then proceed to see whether a line of resistance can be drawn within the middle third. Swain then subdivides the arch and fill into segments, computes their weight and centroid location, and proceeds to draw a line of resistance by the method of drawing a force and funicular polygon. Having satisfied himself that the arch can resist symmetric loading, Swain proposes the investigation of live load on one-half the span. In this case, the funicular polygon is not so easy to draw, the crown thrust being inclined with respect to the horizontal, so Swain employs a procedure for supposing three points of the line of resistance (at the supports and the crown) and drawing a funicular polygon through these three points.

Unsymmetrically Loaded Arch

The process of finding an appropriate thrust line for an unsymmetrically loaded arch is more complicated and can be accomplished in one of two ways. This can be done by a process of finding successive approximations to the thrust line, using the properties of the funicular polygon. Swain (1896) suggests a more exact general procedure for drawing a funicular polygon through any three given points. This makes the trial and error in finding the correct funicular polygon unnecessary, although Swain's procedure is somewhat exacting.

Analysis of Buttresses

Kidder (1886) includes both graphical and semigraphical analysis of the resistance of stone buttresses to thrust. He divides a rectangular buttress into horizontal layers and finds the intersection of the resultant of the thrust and partial buttress weight with the horizontal plane at the bottom of each layer. In analyzing a buttress with steps, such as shown in Figure 13-6, he divides the buttress into trapezoidal segments by means of vertical cutting planes and finds the center of gravity of each trapezoidal segment by the "method of diagonals," a general method for finding the center of gravity of a quadrilateral (Figure 13-7). The diagonals of the quadrilateral are drawn. The length of the shorter segment of each diagonal is set off on the longer side using a divider. Connecting the ends of the transcribed lengths with the dashed line results in a triangle at the center of the quadrilateral. The center of gravity of this triangle (open circle), easily found by medians, is also the center of gravity of the quadrilateral. The centroid of the pier with offsets is found analytically by taking moments of the three sections.

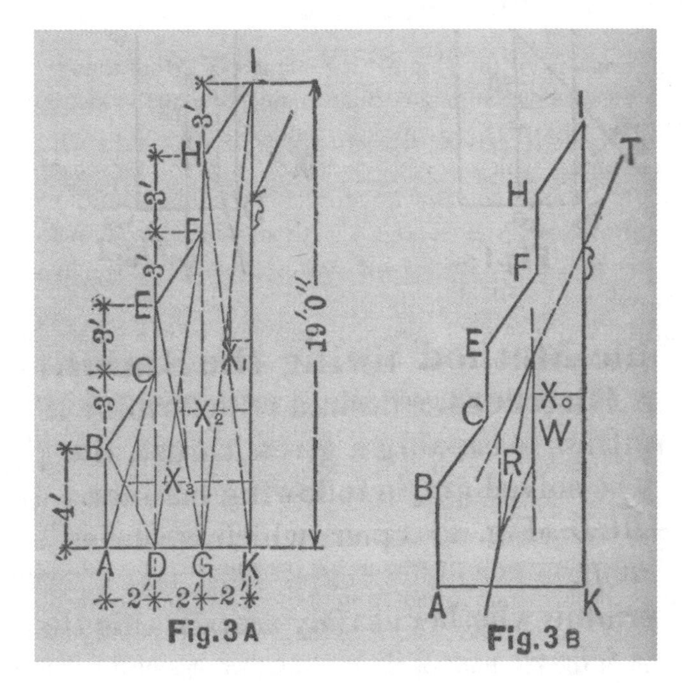

Figure 13-6. Graphical solution of stability of buttresses.
Source: Kidder (1886).

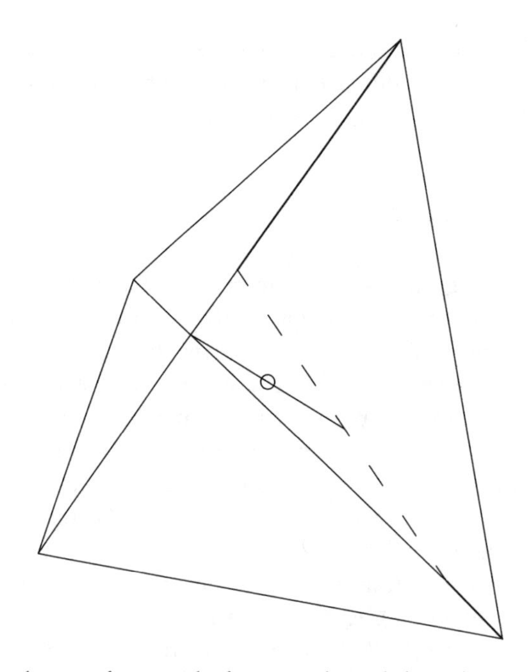

Figure 13-7. Graphical solution of centroid of a general quadrilateral.

The Analysis of Union Arch

The drawing of the analysis of the Union Arch, signed by Montgomery Meigs and Albert Rives, shown in Figure 13-8, is a summary of an engineering analysis of a monumental stone arch (*Souvenir History* 1890). This analysis depends on the application of advanced techniques in graphical statics, which are justified by results from projective geometry. This drawing, dated 1859, is almost coincident with the first publication of Culmann's synthesis of graphical statics. It invokes Méry's (1840) method, which was a method of the analysis of vaults in which the thrust line for a symmetric arch is drawn by the construction of an arbitrary funicular polygon rectified by the procedure as discussed. The method used on the design drawing of the Union Arch begins with drawing a preliminary funicular polygon, which is transformed into a polygon having the required properties of passing through two given points: one at the crown and one at the abutment, and of the crown thrust being horizontal. Based on the arbitrary pole location and starting point, the diagram represents one possible line of thrust for an arch. The drawing in Figure 13-9 illustrates the transformation to the correct polygons, using a simple system of one resultant and two reactions.

When the upper diagram shown in Figure 13-8 is completed on the basis of an arbitrary pole selection and an arbitrary starting point for the funicular polygon, the lower diagram is also determined by the horizontal orientation of the crown thrust, the point through which the crown thrust passes, and the intersection of the crown thrust and the abutment resultant on the line of action of the resultant of the loads. Other legs of the funicular polygon can be filled in by projecting two points onto the line from the upper diagram to the transformed diagram below. This procedure can be justified in terms of the forces represented by the

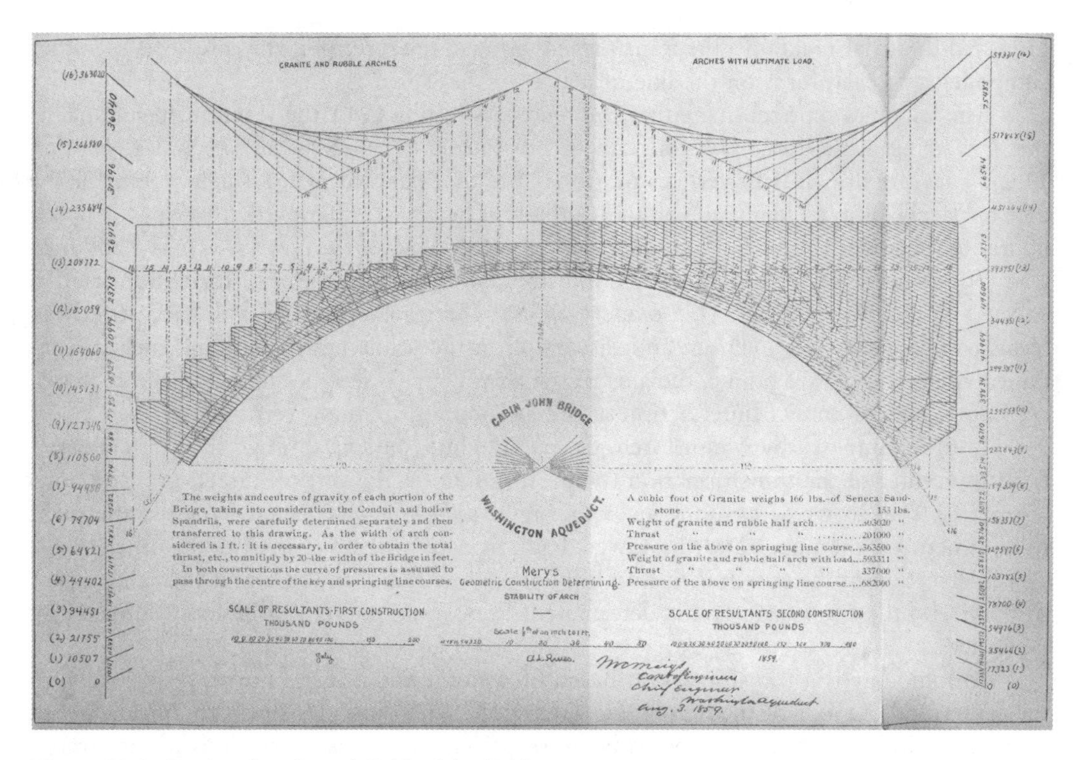

Figure 13-8. Design drawing of Cabin John Bridge.
Source: Unknown (circa 1890).

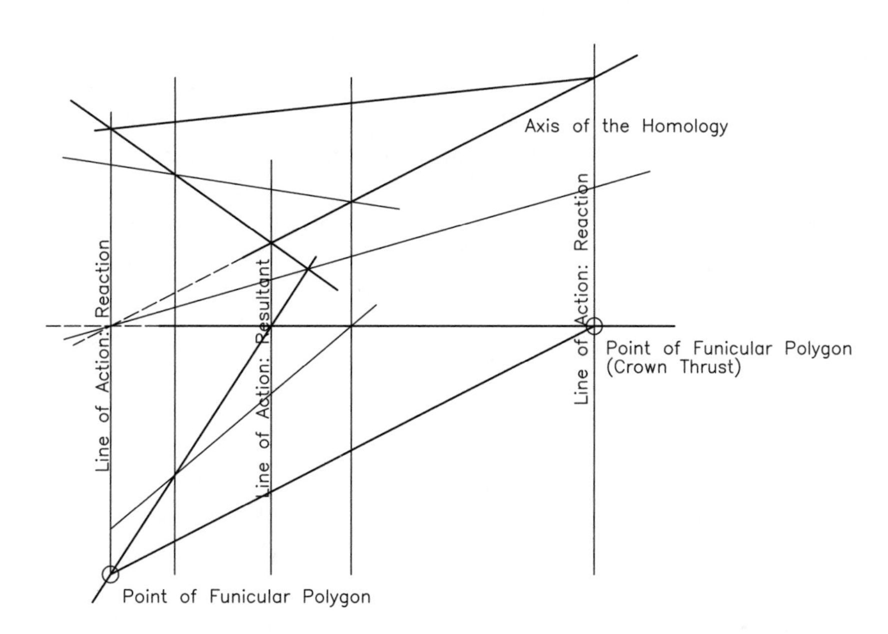

Figure 13-9. Summary of transformation used in Mery's method analysis of Union Arch.

force and funicular diagrams: the transformed string in the force polygon intersects the other components of each force on the line of action of that force.

The triangles representing either the reactions/resultant, or the reactions/resultant for the new string in the force polygon, can be recognized to be Desargues triangles (see Box 11-2). They are drawn with vertices on three lines that pass through a single point, for the case of parallel gravity loads, the point at infinity. The transformation that passes the upper figures to the lower figures is a two-dimensional projectivity of the type known as a collineation (points are transformed to points and lines to lines), and the transformation itself is called a homology by H. S. M. Coxeter (1955). The transformation is defined strictly in terms of incidences (see diagram and discussion in the following paragraph) and is determined when one pair of points, the axis of the homology (Desargues line) and the center of the homology are known (intersection of the line joining common vertices).

In the example of the Union Arch, the parallel lines passing through the vertices determine the center of the transformation (point at infinity in the vertical direction). The horizontal line from crown determines one point on the axis, whereas the line from the abutment to abutment on the upper and the lower diagram determines a second point on the axis. The intersection of the crown thrust with the line of action of the resultant furnishes a pair of points, so the transformation is determined. Remaining points can be found directly by incidences.

A homology is a transformation based on a point and a line, known, respectively, as the center and the axis of the homology (Figure 13-10). Points transform to an image that is collinear with the center and point being transformed. Points on the axis are invariant, that is, their image in the transform is the same as the initial point. The homology is then determined by a pair of a point A and its transform A'. To construct the corresponding transform X' of an arbitrary point X, draw a line through X from the center. The line AX intersects the axis at C. Because the transformation preserves incidences, the line $A'X'$ is also incident with C, while X' must also be incident with OX. Thus, X' is at the intersection of OX and $A'C$.

In the case of the Union Arch, the construction of the homology is used to make the arbitrary thrust line, shown at the top of the drawing, into a thrust line that acts in a horizontal direction at the crown, acts through the abutment of the arch, and is everywhere else

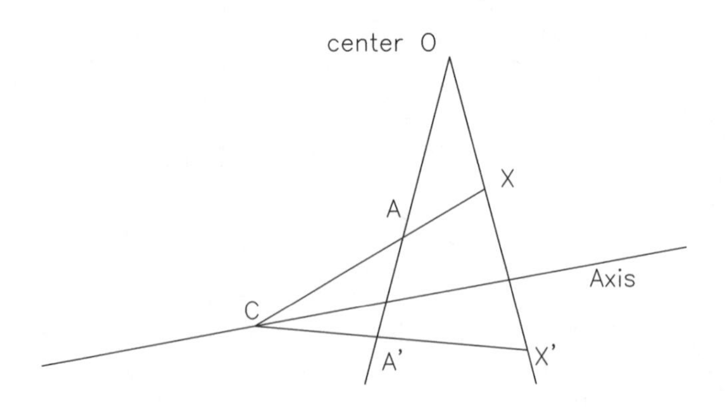

Figure 13-10. Details of a homology.
Source: Adapted from Coxeter (1955).

contained within the arch. Only the last condition is not automatic with this construction: to be satisfied, the arch must have the correct geometry, and the analyst may have to experiment with the construction.

References Cited

Baker, I. O. (1907). *A treatise on masonry construction*, 9th Ed. John Wiley and Sons, New York.

Coxeter, H. S. M. (1955). *The real projective plane*, 2nd Ed. Cambridge University Press, MA.

Haupt, H. E. (1856). *General theory of bridge-construction.* D. Appleton, New York.

Heyman, J. (1982). *The masonry arch.* Ellis Horwood, Chichester, UK.

Howe, M. A. (1906). *Symmetrical masonry arches.* John Wiley and Sons, New York.

Kidder, F. (1886). *The architects' and engineers' pocket-book*, 3rd Ed. John Wiley and Sons, New York.

Méry, M. E. (1840). "Sur l'équilibre des voûtes en berceau." *Annales des ponts et chaussées*, 19, 15–70, and plates 133–134 (in French).

Sondericker, J. (1903). *Graphic statics, with application to trusses, beams, and arches.* John Wiley and Sons, New York.

Souvenir history of Cabin John Bridge. (circa 1890). W. H. Brewton, Washington, DC.

Swain, G. F. (1896). *Notes on the theory of structures*, 2nd Ed. Mimeographed course notes from lectures presented at the Massachusetts Institute of Technology, Department of Civil Engineering, Cambridge, MA.

14

Graphical Analysis of Beams

In the following chapter, we will investigate some effective means of analyzing beams by graphical methods. These methods follow those developed and expounded by Karl Culmann and other workers in graphic statics. Graphical methods were championed by several American authors, including A. Jay Du Bois and Charles Ezra Greene. The methods adopted in Chapter 13 for the analysis of arches are equally applicable to beams and can give a large amount of information in a relatively quick construction. Graphical methods were not as widely used for beams as for arches, because the analytical complexities discussed in Chapter 13 for arches are not necessarily present in beams. Nevertheless, an investigation of the application of graphical methods to the analysis of beams is illuminating.

Beam reactions are found by completing the funicular polygon of the loading and reactions of the beam. Thus, a beam with two concentrated loads *ab* and *bc*, as shown in Figure 14-1, has a force polygon with four strings, *a, b, c,* and *d*. The reactions are designated *ad* and *da*, with the location of the point *d* at some point on the load line, initially unknown. However, because the string *d* must intersect the line of action of the right reaction at *cd* and the left reaction at *ac*, the position of *d* on the force polygon is found by connecting these two points on the funicular polygon and transferring the direction to the force polygon. The location of the resultant of the loads can be found immediately by extending string *a* and *c* to their point of intersection. These procedures are described in every textbook or set of printed course notes on graphic statics, including George F. Swain (1896), A Jay Du Bois (1877), and Mansfield Merriman and Henry Sylvester Jacoby (1894).

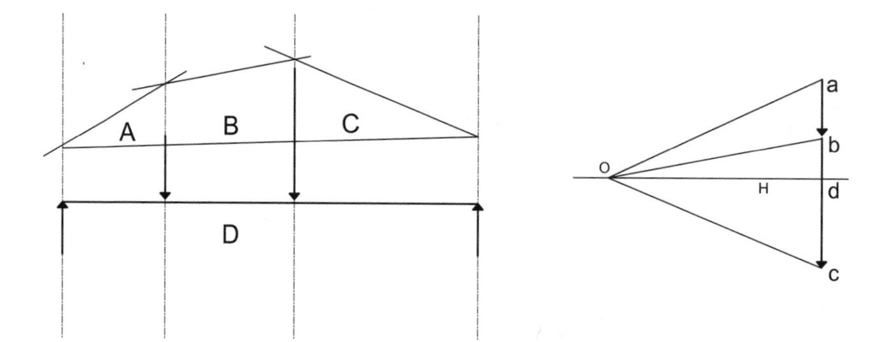

Figure 14-1. Funicular (left) and force (right) polygon for simply supported beam.

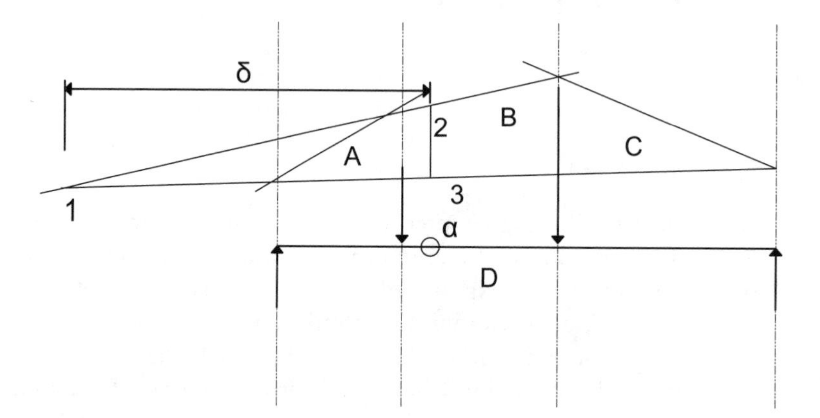

Figure 14-2. Application of Culmann's theorem to determination of bending moments in beams.

Finding Bending Moments in Beams

Finding bending moments in beams graphically is an application of Culmann's theorem. Designating the perpendicular distance from the pole to the load line in Figure 14-1 as H, the bending moment about the point on the beam marked α is equal to the resultant force bd (measured on the force polygon in Figure 14-1) times the horizontal distance δ from α to the resultant at the intersection of strings a and d (measured on the funicular polygon). By inspection of the geometry, it can be seen that the triangle bdO on the force polygon (in Figure 14-1) is similar to the triangle 1-2-3 outlined on the funicular polygon (in Figure 14-2). By proportions, then, the bending moment at α is equal to H times the vertical leg 2–3 of the triangle 1-2-3 (distance between strings b and d). This is a specialization of Culmann's theorem, presented in Chapter 11. The pole distance is measured from the pole to the load line on the force polygon and has units of force, whereas the ordinate of the funicular polygon is measured from the closing string to the funicular polygon along a vertical line and has units of length. As a result of this analysis, the funicular diagram for a beam is equivalent to the bending moment diagram. This result, which is discussed by Du Bois (1877, p. 87), Swain (1896), Charles E. Greene (1877), and others, originates with Culmann (1875).

Graphical statics applies effectively to beams with uniformly distributed loads and with variable loads. For complex variable loads, the application of the method simplifies the determination of the maximum bending moment in a beam. An example of this type of analysis is taken from Du Bois (1877), for the investigation of the moments on a beam due to a complex locomotive loading. In the example, suppose that the maximum bending moment of a 50-ft span beam is to be determined on the basis of the movable load group shown in Figure 14-3. The loading configuration resembles a Mogul engine such as that shown in Figure 14-4. It is necessary to construct a load line for the load group and, using an arbitrary pole location, construct a corresponding funicular diagram. Then, rather than moving the funicular diagram as the loads move, it suffices to move a closing string along the funicular diagram for supports spaced 50 ft apart. The horizontal pole distance remains fixed for all of the variants of this construction. Thus, among the five alternative positions of a 50-ft long closing string shown in Figure 14-3, the position of the load train that

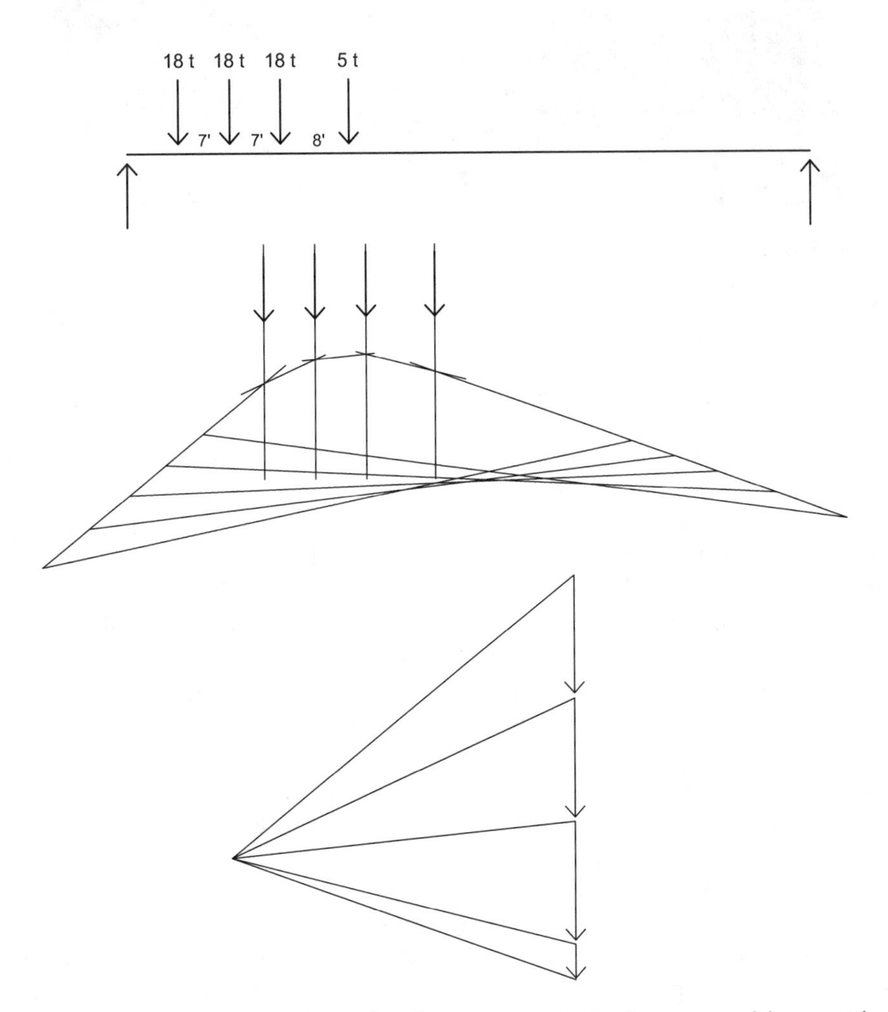

Figure 14-3. Determination of maximum bending moment in simply supported beam with moving load.

Figure 14-4. Mogul engine circa 1900. The cast-iron bridge at Brownsville, PA, also shown in Figure 1-4, is visible in the background of this photograph (under the right-hand arch).
Source: Monongahela Railway Company Photograph Collection, 1893–1993, AIS.2009.07, Archives Service Center, University of Pittsburgh.

produces the largest distance between the funicular polygon and the closing string also produces the greatest bending moment.

Various authors also combined the graphical method of analysis with the principles of beam bending to produce deflection diagrams of beams. This result is made possible by the combination of Culmann's theorem with bending theory, which recognizes the analogy between the calculation of the bending moment from the loads and the calculation of deflections from the curvatures. That is, a beam loaded by the curvature has bending moments equal to the deflection at any point. The method was applied by Fleeming Jenkin (1876) in his article on bridges by determining the bending moment in a beam segment by segment, hence the radius of curvature from $k = M/EI$, then drawing the radii of curvature to scale and scaling the mid-span deflection of the beam (Figure 14-5). Other authors work more formally with the elastic properties of the beam. For instance, James B. Chalmers (1881, p. 199 ff) constructs the elastic curve of a statically determinate beam by loading the beam with the curvature diagram.

As an example of the graphical determination of deflections in a beam, consider the simply supported beam shown in Figure 14-6, which is subjected to a single concentrated load. The first funicular polygon can be used to find the maximum bending moment. Subsequently, this funicular polygon can be subdivided into a series of concentrated curvatures M/EI, and a force polygon can be drawn with pole distance $H' = EI/Han$. In this equation,

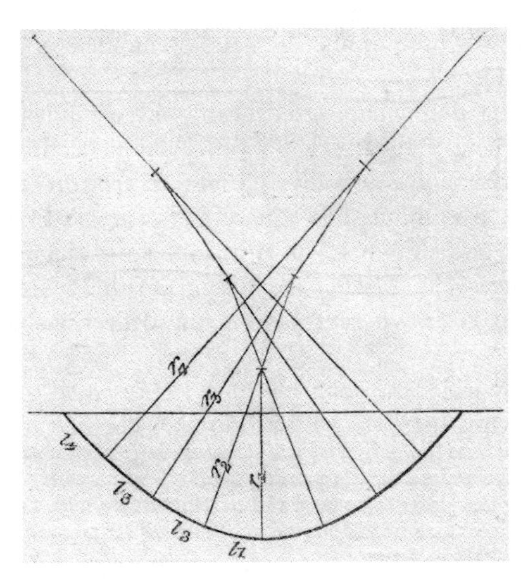

Figure 14-5. Excerpt from Jenkin showing the construction of the elastic curve of a beam based on its curvatures.
Source: Jenkin (1876).

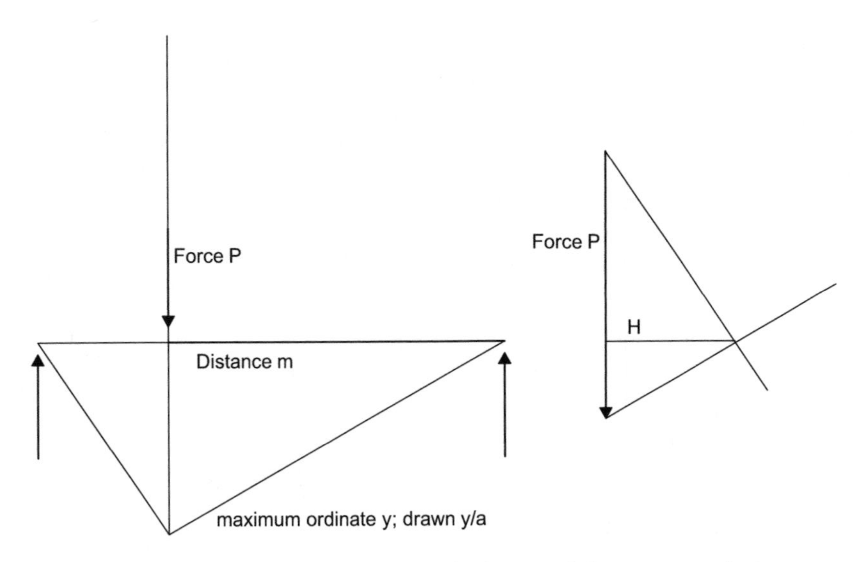

Figure 14-6. First force and funicular polygon in calculation of elastic curve of a beam.

the factor a represents the scale of the funicular polygon representing the bending moments, for example a funicular polygon drawn on a length of 10 in., representing a span of 10 ft, $a = 12$. The factor n represents the diminution of the deflections on the deflection diagram, that is, if the deflections are drawn doubled, $n = 1/2$.

It is possible to take the bending moment diagram for a beam and reapply the calculation of the first funicular polygon to find the bending moment of the curvature, that is, the deflections of the beam. It is necessary to keep track of the units by taking account of the scale of the drawing that was used to find the bending moment and the scale of the drawing used to find the deflections.

Figure 14-6 depicts a simply supported beam with a point load at an arbitrary location. To be able to complete a direct calculation of the deflections, the funicular polygon is drawn. The first funicular polygon is divided into an arbitrary number of segments, and the area of each segment is computed in square feet. The associated force polygon (force polygon 2) is calculated and drawn to an arbitrary scale (Figure 14-7), whereas to the same scale a pole distance $H' = EI/Han$ is chosen. On the basis of this force polygon, a second funicular polygon is drawn representing the deflections of the beam, diminished by the factor n.

In the example shown, let the span be 10 ft (represented on the drawing as 2.5 in.; $a = 48$), the force $P = 7$ kips. Let $EI = 10^5$ in.^{2}-kips. On the first force diagram, $H = 4$ kips,

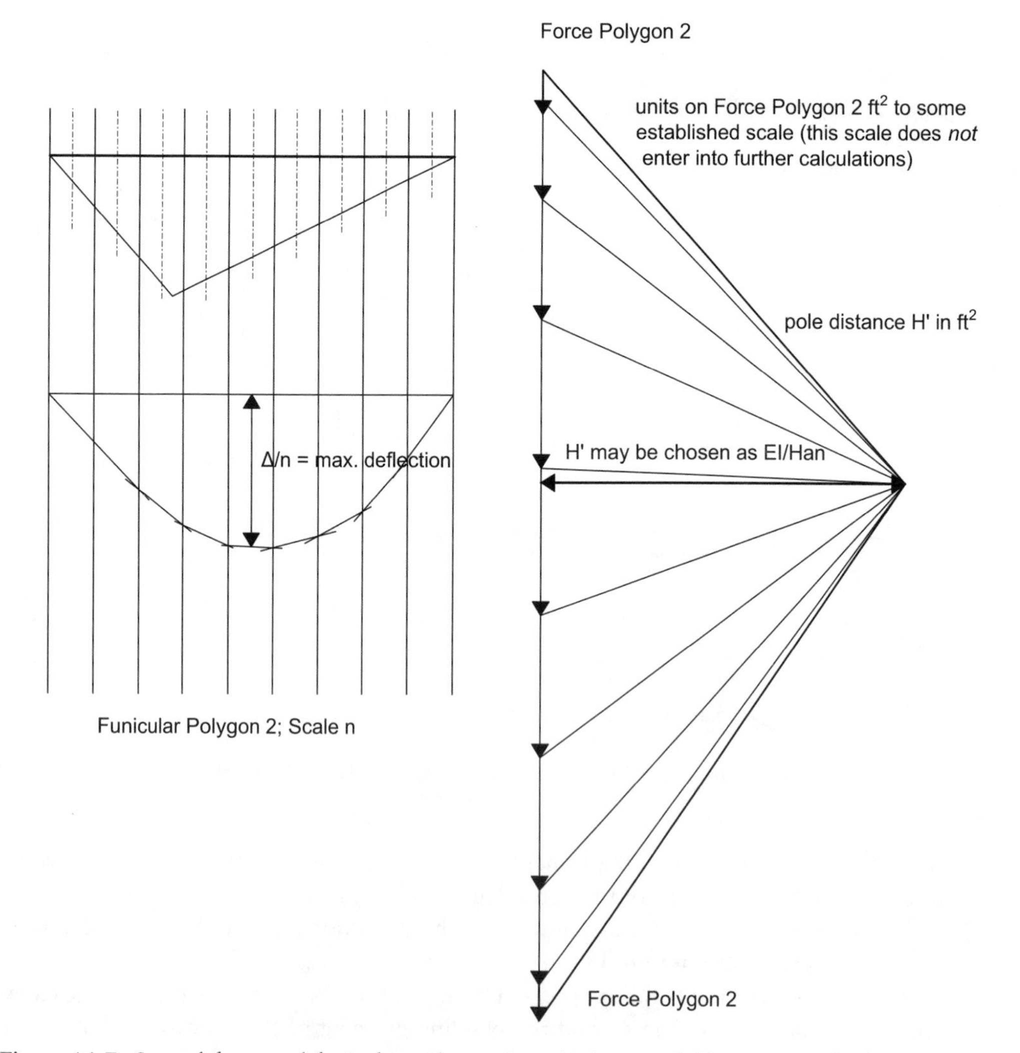

Figure 14-7. Second force and funicular polygon for calculation of elastic curve of a beam.

and the maximum bending moment is 1 in. (ordinate of funicular polygon) × 48 × 4 kips = 176 in.-kips. Then, for $n = 2$, H' is drawn to scale the second pole distance, is 1.8 ft^2. The deflected shape is shown on the deflection diagram (Figure 14-7), and using the factor $n = 2$, can be scaled directly as approximately 2 in.

These discussions make it possible to consider the application of graphic method of drawing the elastic curve to the graphical determination of unknown support moments in continuous span girders. Such methods are discussed by Culmann (1875) and Robert Hudson Graham (1887), among others, in Europe, and in the United States by Du Bois (1877) and Greene (1877). The approaches of Du Bois and Greene differ markedly from each other: Du Bois describes a strictly graphical method of constructing the bending moments at the supports (see Figure 14-8). The method is relatively simple to apply, although challenging to understand, but is sensitive to drafting errors and to accumulating error, and little evidence of the actual use of this method has been found. Greene's more pragmatic method involves the application of formulas to the analytical determination of support moments, followed by graphical construction of the moments between the supports. In speaking of continuous girders, these authors are usually thinking of trussed girders. If these

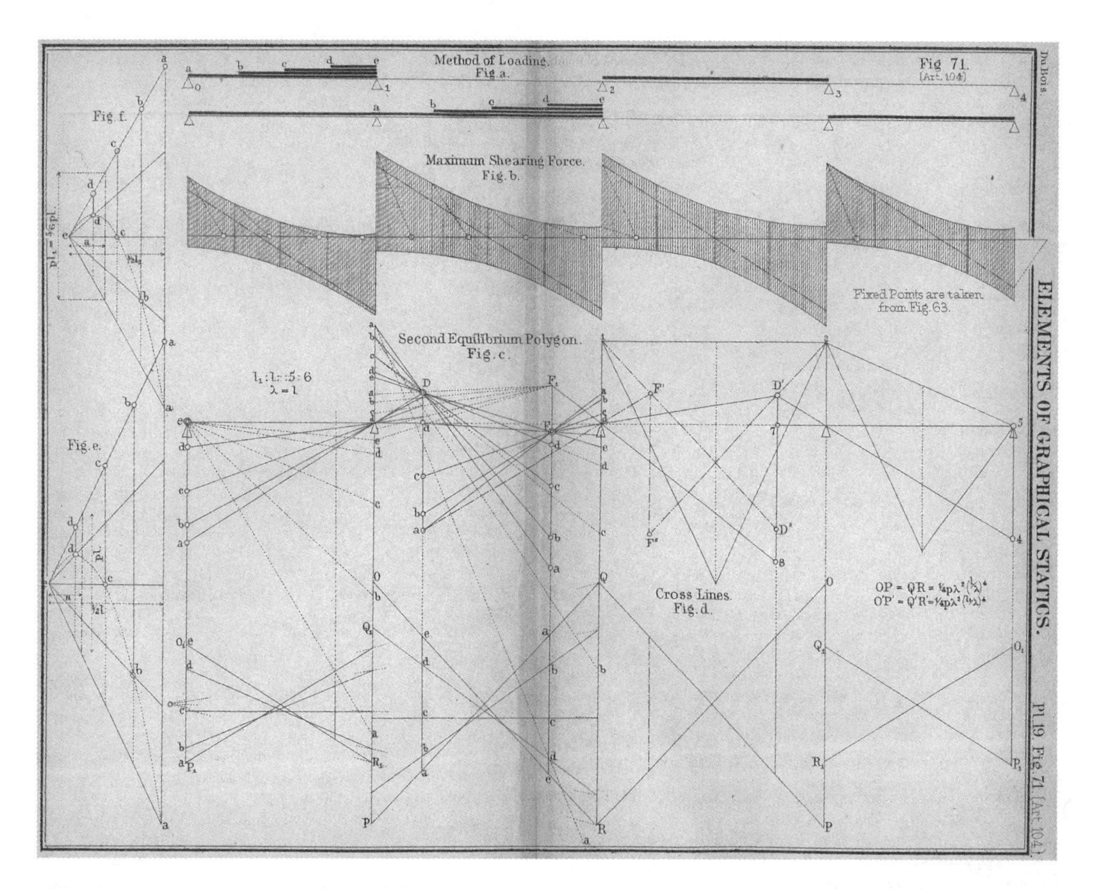

Figure 14-8. Du Bois's method of determining support bending moments in a continuous beam by graphical statics.
Source: Du Bois (1877).

girders are variable in cross section or the top or bottom chords vary along the length, the determination of the bending stiffness for the girders is also approximate.

Greene's Method

Greene presents his graphical method for the analysis of continuous girders, first for a two-span girder and then for a multiple-span girder. Greene uses the moment-area method combined with trial and error as a means of determining deflections with respect to beam tangents. In the two-span girder method, he employs a trial-and-error approach to drawing the closing lines on adjacent spans. For the two-span girder, Greene begins by determining the areas of the moments (or curvatures, if the beam stiffness is variable) for the parts of the bending moment diagram above or below the axis. He observes that if a line tangent to the slope of the beam at a support is drawn, the sum of the moment areas on either side of the interior support has to be proportional to the deflection of the beam end from the tangent line. Thus, the ratio of moment area of *AB* to that of *BC* in Figure 10 of the plate shown in Figure 14-9 must be equal to the ratio of *AM* to *CN*, the line *MN* representing

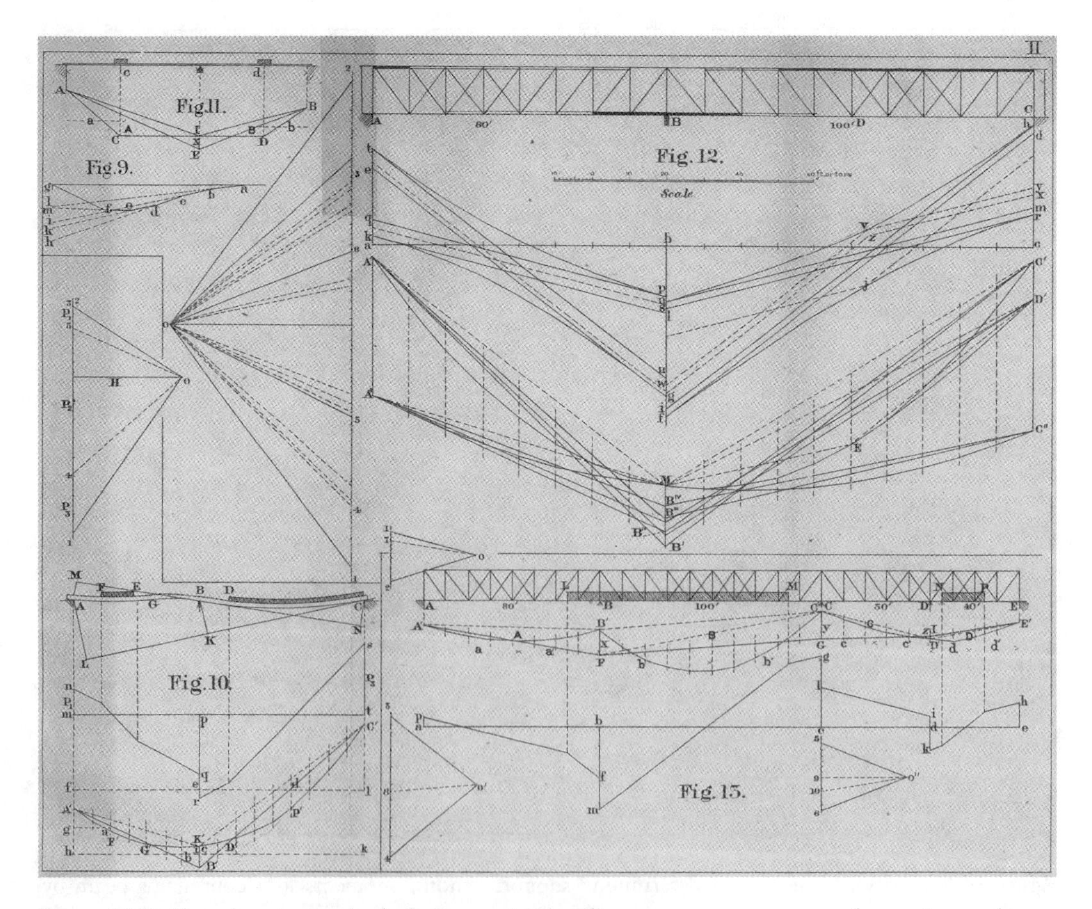

Figure 14-9. Plate showing method of constructing bending moments at support of a continuous beam. Source: Greene (1877).

the tangent to the deflected beam at the interior support. The procedure, then, is to draw a trial closing string *A'B'* and *B'C'* to calculate the moments of the areas of positive and negative moment in each span about the interior support and to see how different they are from equality. According to Greene, no more than one additional trial should be needed to achieve a satisfactory solution by this method. However, it is also possible to superimpose the effect of the curvature of a simply supported beam, which can be computed for any state of beam loading, with the moment area of the triangle representing the bending moment induced by the support moment, with the height of the triangle as an unknown and its centroid as an invariant. Then the determination of the support moments for arbitrary loading reduces to an algebraic equation determined on the basis of quantities measured on the diagram. Although this is properly categorized a semigraphical method, as soon as the support moments are determined, the remainder of the moments can be calculated graphically for the loading condition under investigation. This method is applied to a two-span beam in Box 14-1.

Box 14-1. Example Application of Greene's Method

We will use the example of a two-span beam with two different span lengths and with the live load placed in the longer span only, to demonstrate the application of the semigraphical version of Greene's method. The beam has an interior support, designated *B*, a left span *AB* equal to 80 ft, and a right span *AC* equal to 100 ft. Dead load is 1,000 lbs/ft, and live load is 2,000 lbs/ft. The entire live load will be placed on the longer span, and the live load will be removed from the shorter span to obtain the greatest possible positive bending moment. Then the ratio of the deflection at *A* from the tangent at *C*, the line *AL* to the deflection at *B*, the line *BK*, using the notation of Greene's Figure 10 (Figure 14-9), has to be equal to 1.8, the ratio of the combined spans to the longer span. Based on the second moment-area theorem, this is equal to the moments of the areas of the entire system about *A* to the moment of the moment area of the shorter span about point *B*. Given a support moment equal to *f*, this reduces to

$$\begin{aligned}&0.0833(1\text{ kip/ft})l_1^3(0.5\,l_1)+0.0833\,(3\text{ kip/ft})l_2^3(l_1+0.5l_2)\\&\quad-0.5\,f\,l_1(0.667\,l_1)-0.5\,f\,l_2(l_1+0.333\,l_2)\\&\quad=1.8\left[0.0833(3\text{ kip/ft})l_2^3(0.5l_2)-0.5\,f\,l_2(0.333l_2)\right]\end{aligned}$$

substituting the specific relation $l_1 = 0.8\ l_2$ for this problem, we obtain

$$\begin{aligned}&0.342\,l_2^4\text{ kips/ft}-0.780\,f\,l_2^2\\&\quad=1.80\,(0.125\,l_2^4\text{ kips/ft}-0.167\,f\,l_2^2)\end{aligned}$$

from which we obtain

$$f=0.244l_2^2\text{ ft}^{-1}\text{kips}$$

$$f=2440\text{ ft-kips}$$

When this quantity is determined, it can be divided by the pole distance (force units) and laid off on the funicular diagram of the two-span beam. The remainder of the moments in the beam then can be determined by scaling from the moment diagrams, such as in Figure B14-1-1.

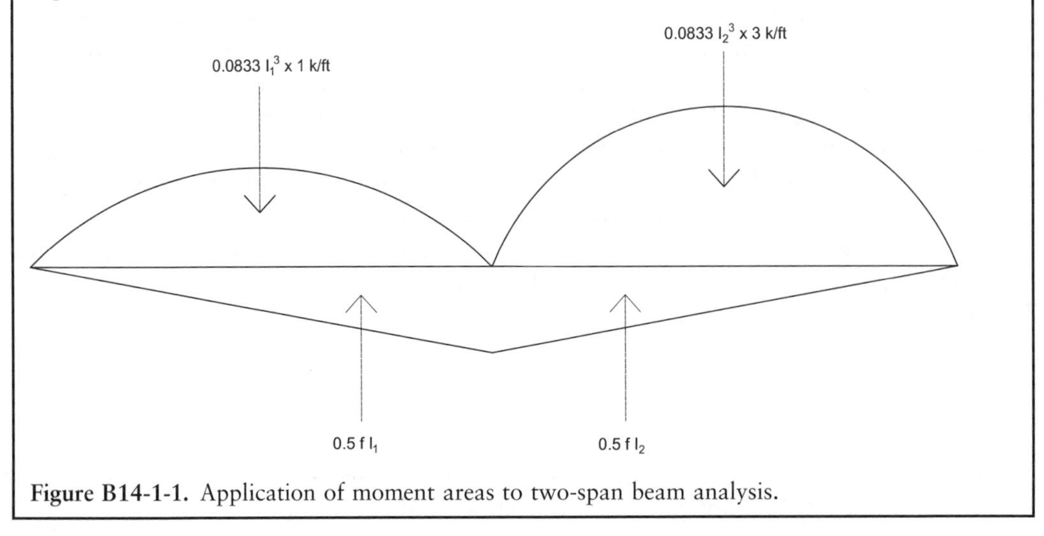

Figure B14-1-1. Application of moment areas to two-span beam analysis.

The application of graphical methods to the analysis of multispan beams, or to the determination of the deflections of beams, has potential for avoiding tedious and error-prone calculations. The result of such graphical calculations is a set of diagrams that convey the important information about a beam at a glance that have their own quality of elegance and completeness. However, as a practical method for beam analysis, these methods appear to have been little used in the United States in the design of continuous beams. Instead, approximate methods or long tabular calculations or, especially the use of simply supported girders in place of continuous girders, found greater favor.

References Cited

Chalmers, J. B. (1881). *Graphical determination of forces in engineering structures.* MacMillan, London.

Culmann, K. (1875). *Die graphische statik.* Meyer and Zeller, Zürich.

Du Bois, A. J. (1877). *The elements of graphical statics.* John Wiley and Sons, New York.

Graham, R. H. (1887). *Graphic and analytic statics.* Van Nostrand, New York.

Greene, C. E. (1877). *Graphical method for the analysis of bridge trusses.* John Wiley and Sons, New York.

Jenkin, F. (1876). *Bridges: An elementary treatise on their construction and history.* Adam and Charles Black, Edinburgh.

Merriman, M., and Jacoby, H. S. (1894). *A text book on roofs and bridges, part 2*, 2nd Ed. John Wiley and Sons, New York.

Swain, G. F. (1896). *Notes on the theory of structures*, 2nd Ed. Mimeographed course notes from lectures presented at the Massachusetts Institute of Technology, Department of Civil Engineering; published in book form (1927) in *Structural engineering: Stresses, graphical statics, and masonry*. McGraw-Hill, New York.

15

Graphical Analysis of Portal Frames and Other Indeterminate Frames

The determination of forces in portal frames, described analytically in Chapter 10, can also be accomplished graphically. The graphic method extends easily to the analysis of column-supported truss roofs in mill buildings, stabilized by knee braces. In addition, other authors touch on further applications of graphic statics to the analysis of rigid frames and other statically indeterminate structures. These topics form the following chapter, primarily using Milo Ketchum's (1903) comprehensive treatment of the analysis of portal frames by graphical methods. Because the portal frame is one degree statically indeterminate, it is necessary to determine, by solution or by assumption, one redundant quantity before the structure can be analyzed. The standard assumption, still used by present-day engineers, for a one bay-one story portal frame is that the horizontal reaction is evenly divided between the two columns. Where there are more columns, some other assumption must be made regarding the distribution of horizontal forces to the columns. The algebraic method of portal analysis in Chapter 10 also uses this assumption. Jerome Sondericker (1904) makes a more accurate assumption that the deflections at the top of the windward and leeward columns are equal, which is especially applicable to frames where the windward columns receive lateral load directly.

Ketchum's method for the graphical analysis of a portal frame requires the insertion of fictitious truss members so that the structure can be analyzed as a truss. The portal of type (a), presented in Chapter 10, Figure 10-3, is shown in Figure 15-1, along with the fictitious truss members (shown dashed). The three web members of the portal, found be to 0-force, are drawn lightly. For total lateral force of 2,000 lbs, and dimensions h = 24 ft, d = 16 ft, and s = 16 ft, the force diagram

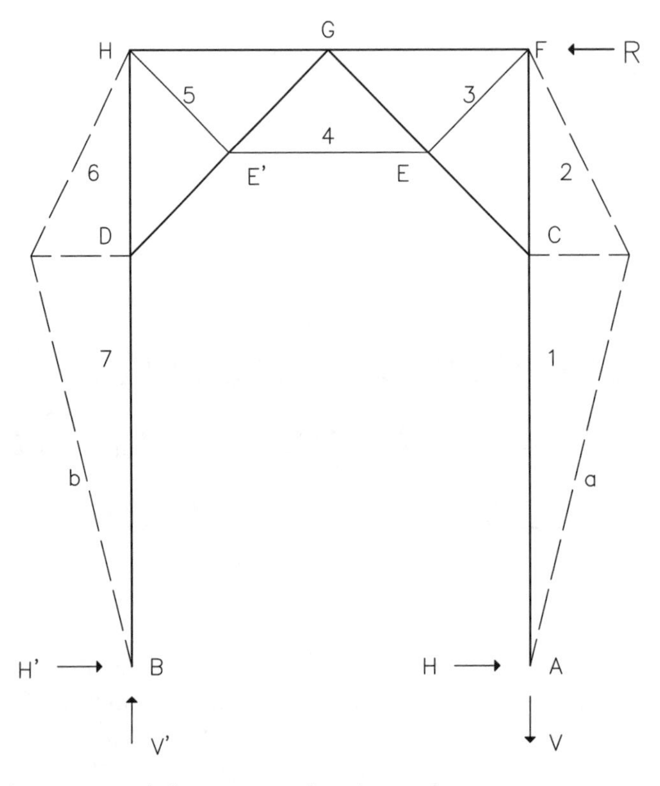

Figure 15-1. Portal type (a), with fictitious members inserted.
Source: Ketchum (1903).

that results is shown in Figure 15-2. On the force diagram, the fictitious members are also shown dashed. The truss is numbered according to Bow's (1873) notation with exterior zones *a* and *b*, and interior zones 1–7.

In the diagram in Figure 15-2, the forces in the fictitious elements are shown as dashed lines, whereas the forces in the actual elements can be read off the diagram. The reactions 4-*a* and *b*-4 are shown as solid lines. The force in the portal strut 3-*a* is found to be 2,000 lbs, and the force in the portal tie 3-4 is 3,000 lbs × $\sqrt{2}$. Because the fictitious forces in 2-*a* and *b*-6 are substitutes for the shear in the column, the actual position of numbers 2 and 6 on the force diagram is on the vertical projection to the load line (2′ and 6′), indicating that 2-3 and 5-6 have 0 axial force. Similarly, from the force 7-4 on the diagram must be subtracted the vertical component of 7-*a*, resulting in the projection of 7 to the load line. The force in 7-4 is thus 3,000 lbs. Similar diagrams for Ketchum's portals (b) through (f) are shown in the Box 15-1.

The result of these analyses is that the force diagrams of the bars in the portal frame determine a regular pattern, which is related to the pattern of the bars being analyzed and which can be constructed quickly for the determination of the forces in these bars. As Ketchum also shows, a similar procedure can be used to solve for the wind load forces in a mill building bent, consisting of a truss, a pair of columns, and knee braces extending from near the top of the column to the first bottom chord panel point in the truss, a form of construction that was widespread in the late nineteenth century.

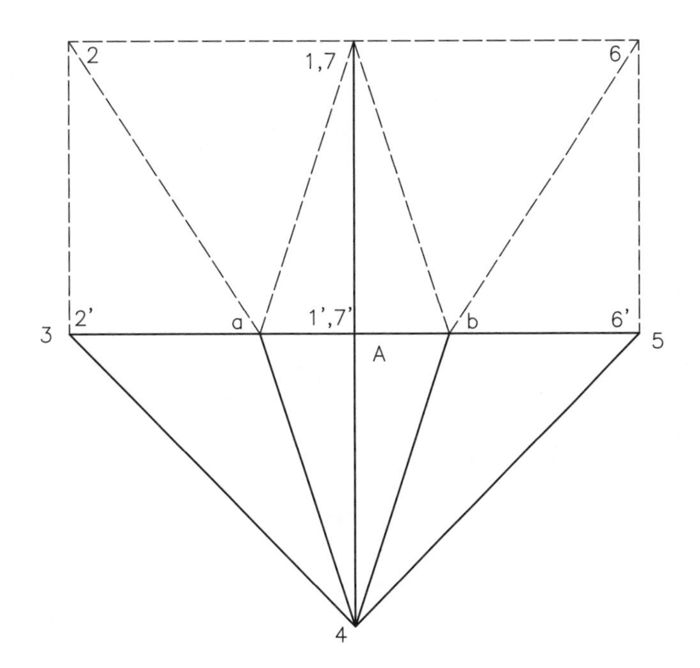

Figure 15-2. Force diagram for portal type (a).
Source: Ketchum (1903).

Box 15-1

Ketchum applies: a procedure similar to the one described in the body of this chapter to portal (b) (see Figure 10-3). Figure B15-1-1 is Ketchum's representation of the graphical analysis of this portal. This system is considered similar to a three-hinged arch, as long as there is a joint in the middle of the strut (the dashed lines represent the lines of action of the resultants, passing through the middle of the top chord). The analysis proceeds normally through the establishment of points *d*, *a*, and *b* representing the external reactions. On this basis, all of the points shown in the figure can be established, except the solid lines 2-3 and 12-11. To see how to obtain the actual force in 11-*d*, for instance, it is necessary to recognize that the force in this member is the vertical component of the sum of the fictitious forces 11-12 and 12-*b*. This is found on the diagram by dropping verticals from points 3 and 11 to the load line, which is the procedure given by Ketchum.

Portal (c) is illustrated in Figure B15-1-2, while Figure B15-1-3 shows the graphic analysis of portal (c). Using the fictitious members the portal frame (c) can be solved as indicated in Figure B15-1-2. Similarly to the other portals described, when the fictitious members are removed, the point 1 has be projected to the point 1′ on the load line—the physical meaning of this is that the vertical component of the force in 1-*b*, a fictitious member, is deducted from the force in C-1. Similarly, the points 2 and 9 are projected to 2′ and 9′. The axial force in C-1 is then correct on the diagram, although no account is taken of the shear or bending moment in this member. The forces in the remaining bars of the frame are given on the diagram in Figure B15-1-3.

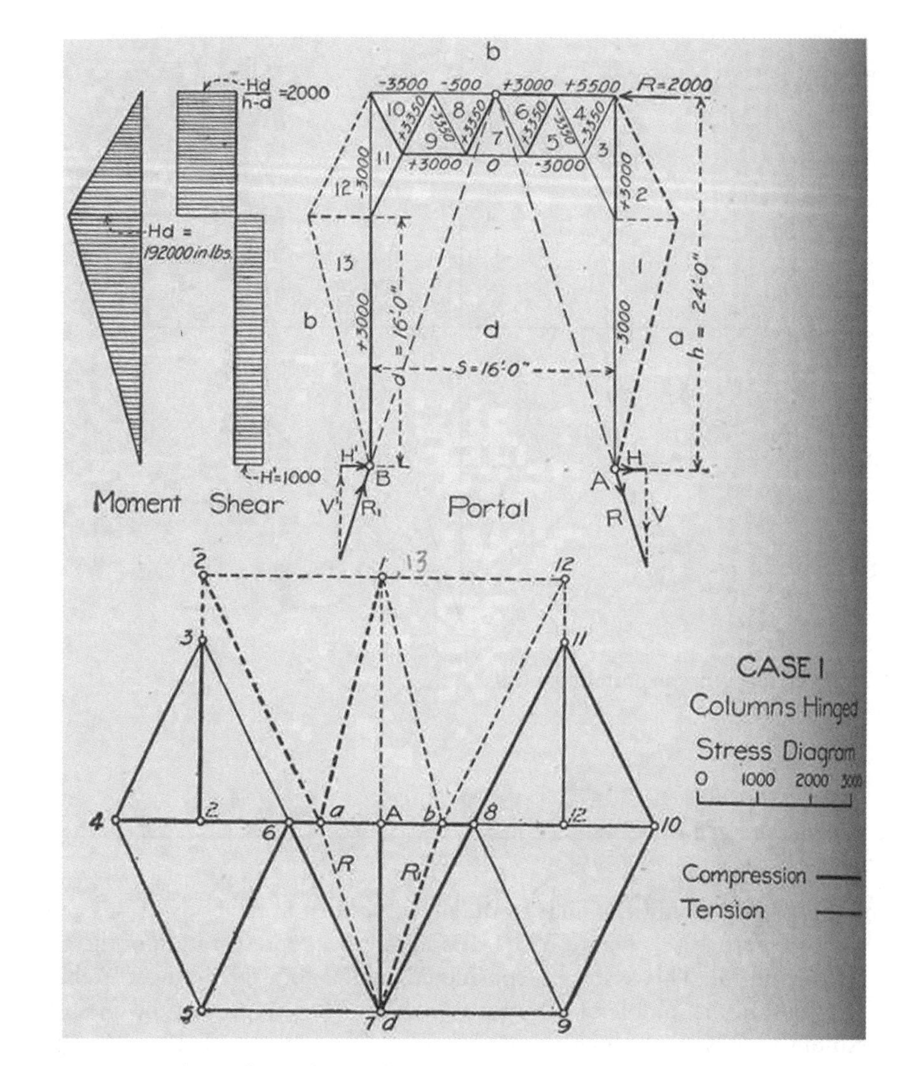

Figure B15-1-1. Analysis of portal type (b).
Source: Ketchum (1903).

To modify this discussion for the portal c′ (Figure 10-3), where the diagonals act both in tension and compression, it is necessary to add panels to the portal. This is shown in Figure B15-1-4. The vertical reaction at the leeward column is equal to the vertical and horizontal shear in each panel of the portal frame, because the panels are square ($\theta = 45°$). The vertical component of V is divided equally between each strut/tie in each panel, resulting in the diagram shown in Figure B15-1-5.

Portal type (d) is illustrated in Figure B15-1-6, while its force diagram is shown in Figure B15-1-7.

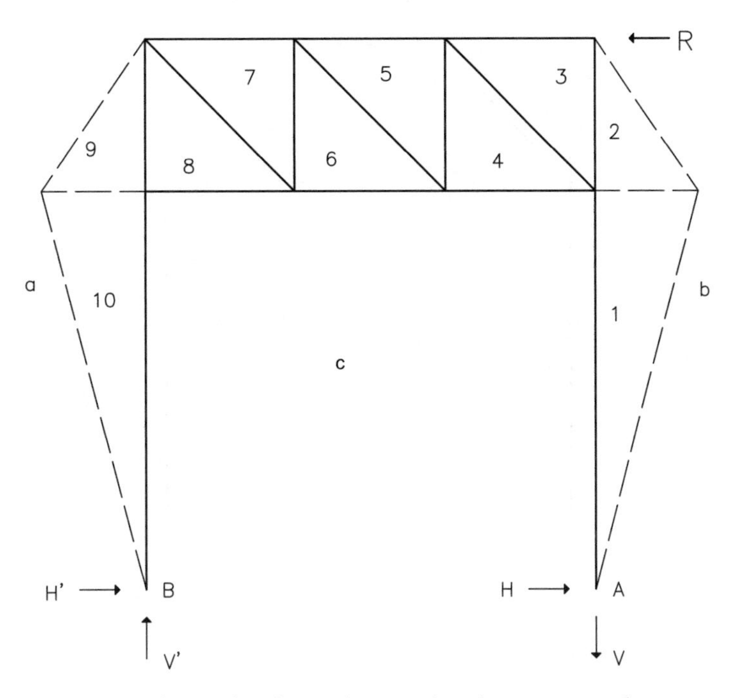

Figure B15-1-2. Portal (c): diagonal web members analyzed in tension only.

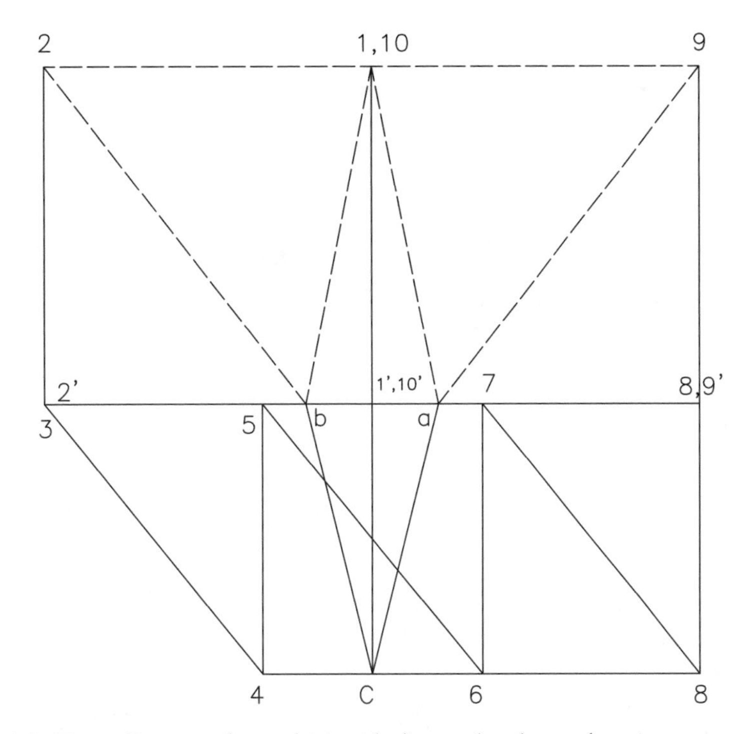

Figure B15-1-3. Force diagram of portal (c) with diagonal web members in tension only.

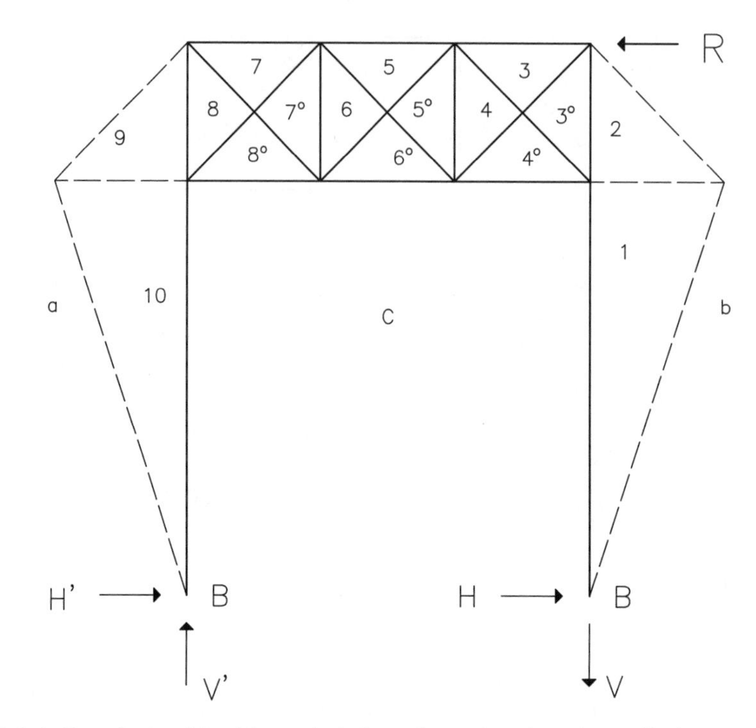

Figure B15-1-4. Portal type (c) with vertical shear force distributed equally between tension and compression web members in a panel.

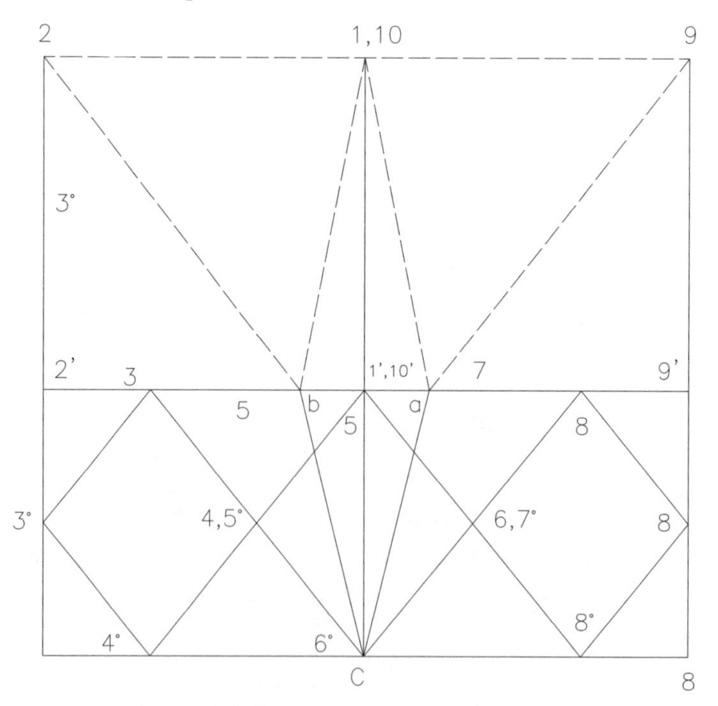

Figure B15-1-5. Force diagram of portal type (c) with vertical shear force distributed equally between tension and compression web members in a panel.

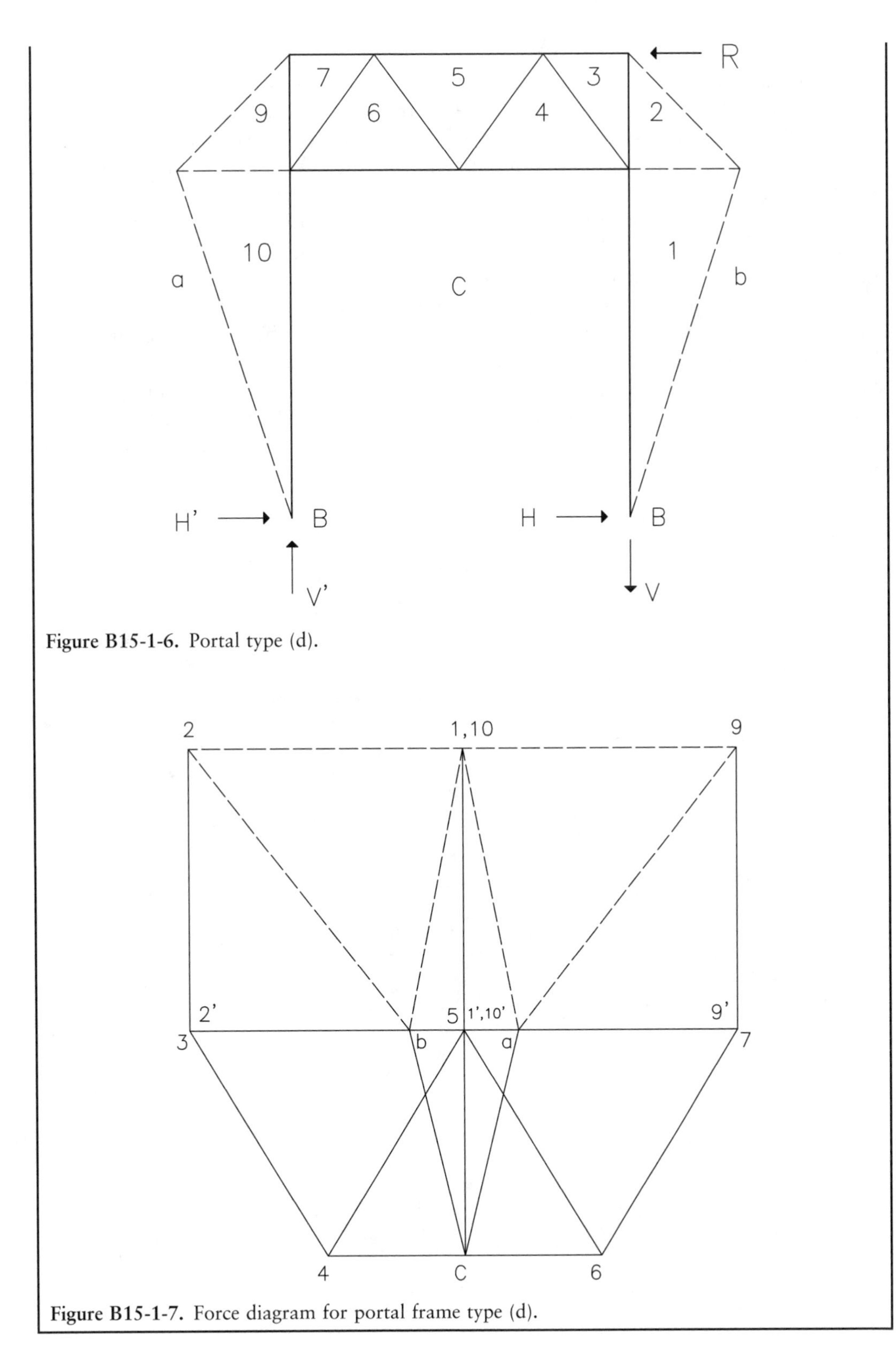

Figure B15-1-6. Portal type (d).

Figure B15-1-7. Force diagram for portal frame type (d).

Ketchum considers two primary types of bents: those that have pinned column bases and those that have fixed column bases. For the pinned base bent, the force diagram is shown in Figure 15-3: external forces produce the triangle *ABC*. The fictitious members *ab*, *bc*, and *cB* are established, and the internal forces are solved by standard procedures of graphic statics. The final position of point *c*, removing the effect of the fictitious members, is at the intersection of the load line and 1-*c*, that is from *c*-1 as shown must be deducted the vertical component of the fictitious force *c*-*x*. This point is located on Ketchum's solution of the bent shown in Figure 15-3. In constructing this diagram, Ketchum uses *x* for each of the spaces between purlin loads, rather than a separate letter for each space.

The analysis for a system with fixed column bases is similar to the previous analysis, except that the point of inflection in the column is chosen to be half the distance from the ground to the knee brace connection. From there, the structure is divided, the column shear is applied at the half height of the column, and the analysis proceeds similarly to the analysis as previously outlined. An example of a structure considered to have fixed column bases is furnished by Berlin Iron Bridge Company's Newport News Shipbuilding Dry Dock, shown in Figure 15-4.

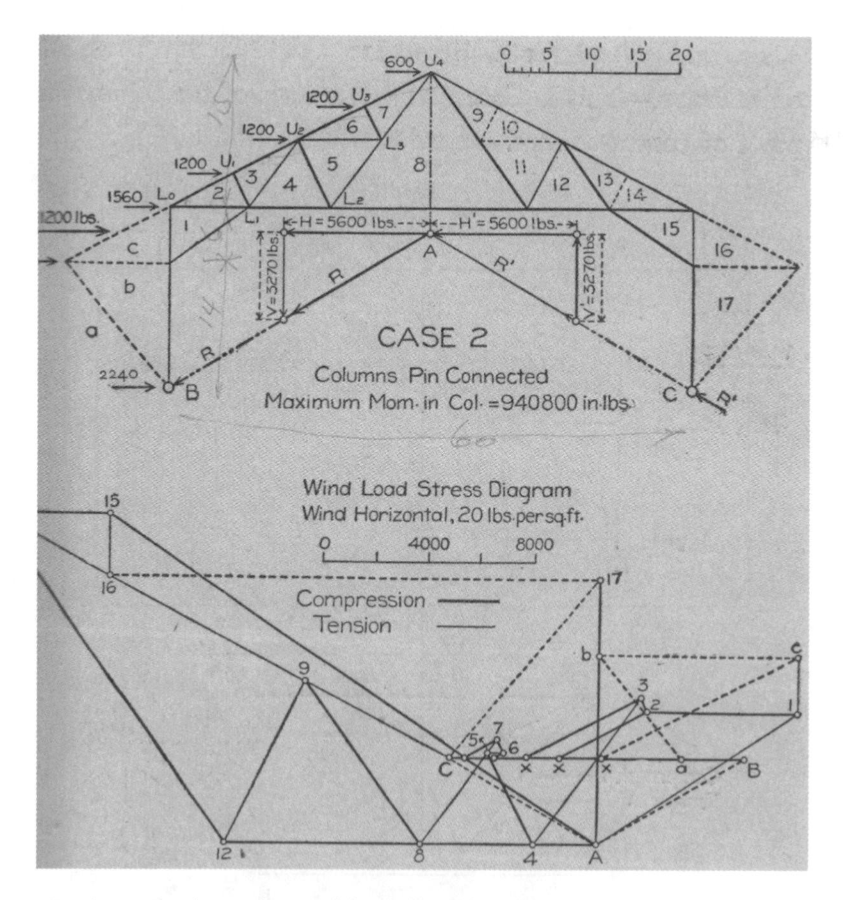

Figure 15-3. Ketchum's analysis of a wind-loaded mill building bent.
Source: Ketchum (1903).

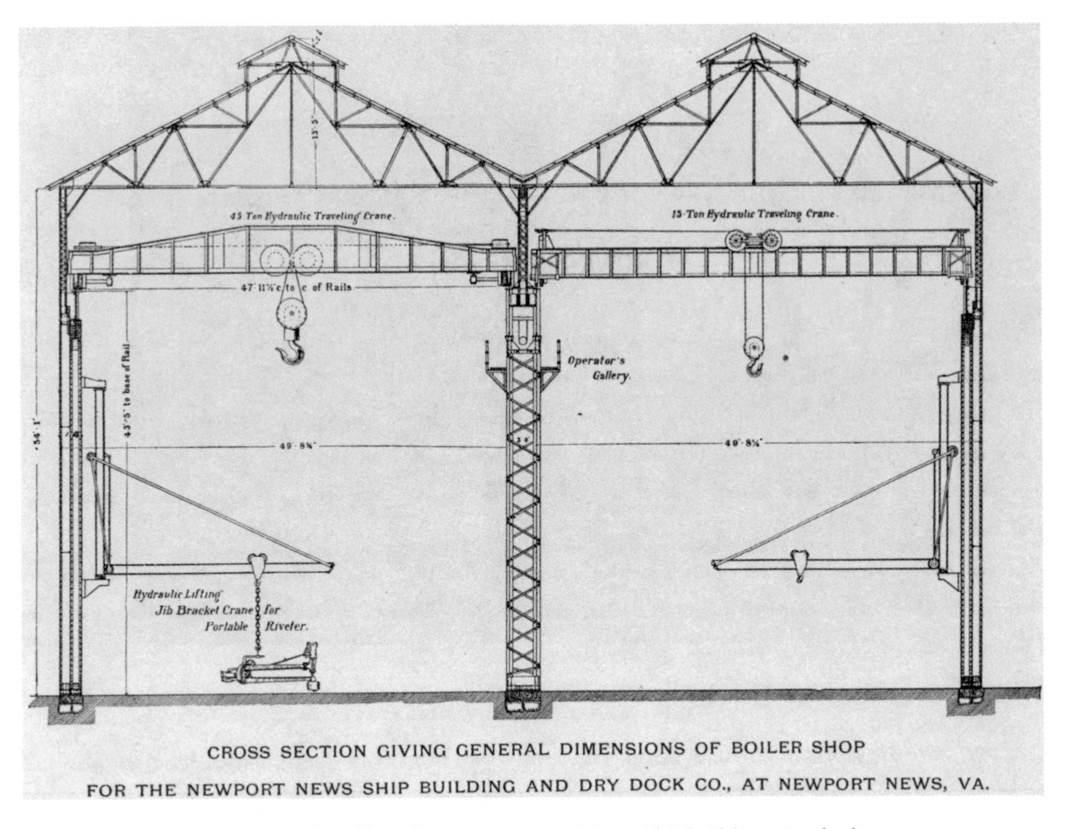

Figure 15-4. Fixed base mill building bents. Newport News Shipbuilding Drydock.
Source: Berlin Iron Bridge Company (ca. 1890).

Ketchum gives similar attention to a multiple portal. In this case, he divides the lateral force equally among the columns, whereas the vertical reaction is in proportion to the distance from the center of the multiple bent. Thus, for a five-bay, six-column portal (shown in Ketchum 1903, Figure 63) the horizontal force at the base of each column is one-sixth of the total horizontal force, whereas the vertical reaction at the exterior column is five times the vertical reaction at the column supporting the middle bay, based on the ratio of distances to the center of the system. Although no assumption is made for the location of the point of inflection in the portal strut, the previous assumptions ensure that the point of inflection is located at the mid-span of each strut. The compound system of portals providing lateral support to the train shed in the Harrisburg, PA, train station, constructed in 1887, provides an example of a multibay portal, encompassing 20 bays in all (Figure 15-5).

As an approximate method, Sondericker (1904) proposes resolving the wind force on the windward column into a force applied at the top of the column, a force applied at the level of the knee brace, and a force applied at the base, similar to Ketchum's procedure. Sondericker advocates a semigraphical approach in which the forces on the columns are found independently of the graphic analysis of the truss. The method is described in detail in Box 15-2. In the case presented there, Sondericker's method produces considerably larger

Figure 15-5. Train shed at Harrisburg, PA, train station.

Box 15-2

Sondericker (1904) shows a different graphic method for the analysis of mill building bents, which is subsequently used by Arthur Shumway (1904) in his bachelor of engineering thesis. Sondericker's work presents various clever graphic statics solutions to various problems of combined bending and axial tension/compression. The solution Sondericker, and later Shumway, employs for a mill building bent with knee braces is to divide each column into a part that resists axial forces only and into a part that acts as a beam, the two parts connected by rigid links. Analytical statics is employed to determine the reactions and the internal forces in the beam, whereas graphical statics is applied to the columns and the trusses that they support. This procedure is illustrated in the diagram from Sondericker (pp. 82–84) and in the example that follows. Ketcham's and Sondericker's procedures have identical results where no force is considered to act directly on the windward column. For other load distributions, the results of Sondericker's method differ from the results obtained by Ketchum's method.

For comparison and illustration a mill building bent with both ends of the columns pinned (illustrated in Figure B15-2-1) will be considered. Given normal loading at each windward panel point of 4 kips (one-half at eave and crown) and a value of W of 400 lbs/ft, $c = 20$ ft, $a = 15$ ft, panel lengths of 10 ft (40 ft overall), and a roof pitch of 6:12, this structure can be investigated by Sondericker's and Ketchum's methods. In Ketchum's method, the total horizontal force of 11,580 lbs is divided equally between the two horizontal reactions, $H_1 = H_2 = 5{,}790$, $V_1 = -1{,}130$ lbs, $V_2 = 6{,}030$ lbs, $S_1 = 7{,}160$ lbs, $S_2 = 23{,}160$ lbs $R_1 = 9{,}370$ lbs, $R_2 = 17{,}370$. To find these quantities graphically by Ketchum's

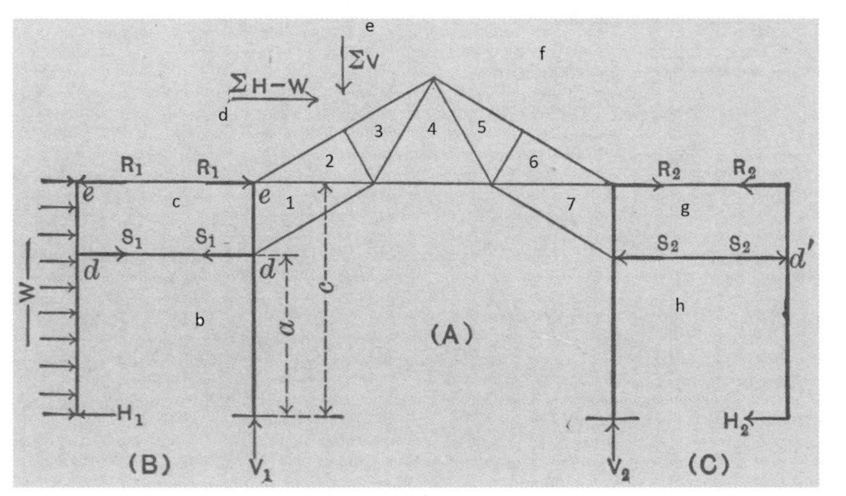

(a)

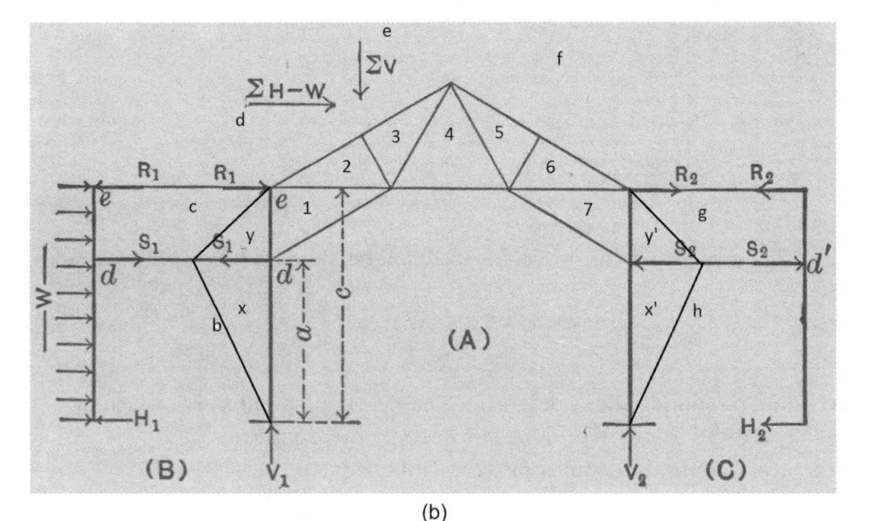

(b)

Figure B15-2-1. Stresses in a mill building bent (a) by Sondericker's method and (b) by Ketchum's method.
Source: Modified from Sondericker (1904).

method, the diagram shown in Figure B15-2-2 is constructed. In working with the loads on the windward column, Ketchum applies this load at the top at the level of the knee brace and at the base, with the distribution based on tributary area.

Based on the diagram in Figure B15-2-2, the force in the knee braces 1-*a* and *a*-7 can be found to be 7,300 (horizontal component 6,500) and 25,600 (horizontal component 23,000). The windward column *x*-*a* has a force on the diagram of about 4,600 lbs from which the vertical component of about 6,000 lbs of the fictitious member *b*-*x* must be deducted, yielding a compressive force of 1,400 lbs. Similarly, the leeward column is found to have a force of approximately 6,400 lbs compression.

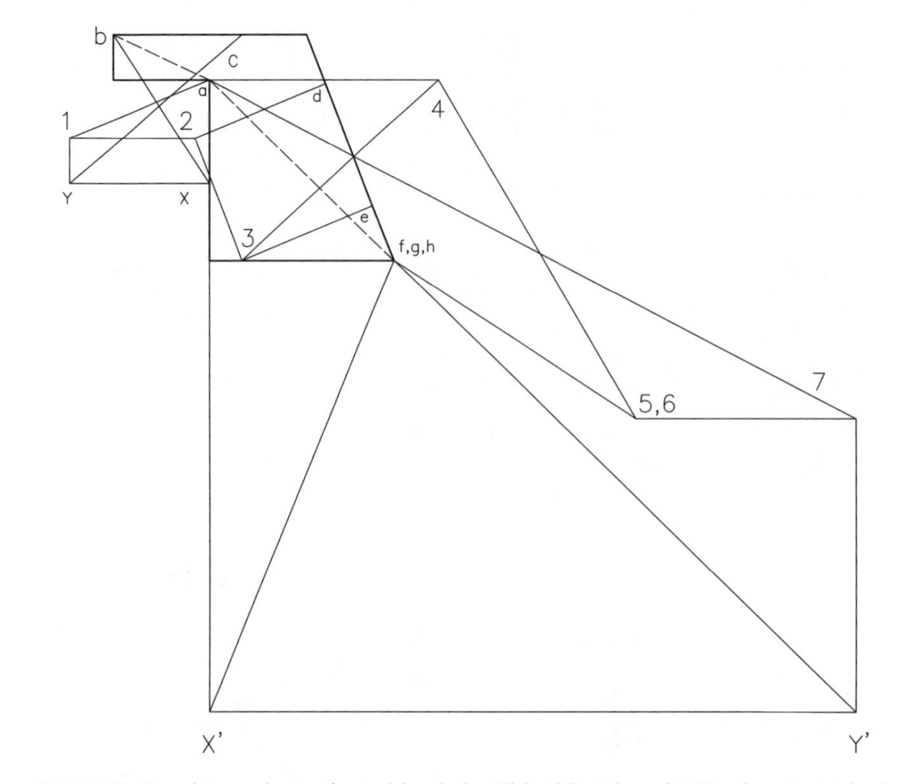

Figure B15-2-2. Graphic analysis of wind-loaded mill building bent by Ketchum's method.

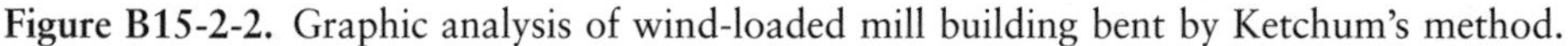

According to Sondericker's derivation, based on equal lateral deflections of the truss, the following are the calculated values of each of the quantities.

In the aforementioned equations, ΣH is inclusive of W.

$$S_1 = 23,200 - 11,200 = 12,000$$

$$S_2 = 23,200 - 4,800 = 18,400$$

$$H_1 = 5,800 + 1,200 = 7,000$$

$$H_2 = 5,800 - 1,200 - 4,600$$

$$R_1 = 17,400 - 4,400 = 13,000$$

$$R_2 = 17,400 - 3,600 = 13,800$$

The formulas used in determining these values are given by Sondericker (1904, p. 86) for the case where the column bases are pinned. In consequence of these values and the removal of the force W from the trussed system

$$V_1 = 1{,}150 \text{ (upward reaction)}$$

$$V_2 = 6{,}050$$

These forces are then inserted into the graphical analysis of the roof truss and the knee braces, resulting in the force diagram shown in Figure B15-2-3. Sondericker's method gives a greater force to the windward knee brace and a lesser force to the leeward knee brace.

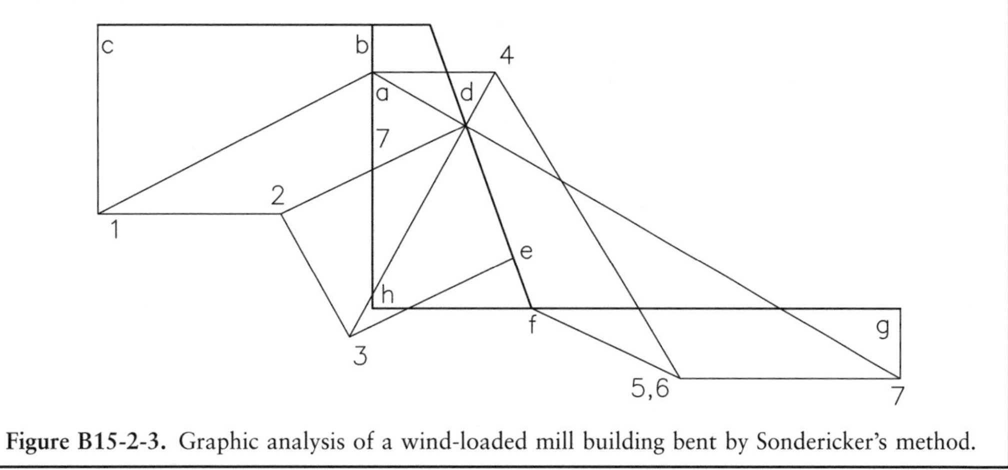

Figure B15-2-3. Graphic analysis of a wind-loaded mill building bent by Sondericker's method.

forces in the knee braces. However, if no force is applied directly to the windward column, then Sondericker's and Ketchum's methods yield the same results.

In general, the selection and the design of a lateral system for a bridge or a building was recognized as an important engineering decision. The literature on the analysis and design of bridges includes extensive discussions of the systems for transferring wind forces from the span of the bridge to the supports, and the discussion of the design of the portal itself is significant. The varied forms of portals are a reaction to the difficulties of obtaining moment resistance at the joint between the portal strut and the column. Similar problems existed in the design of industrial buildings, which, when stripped of the masonry walls that generally provided lateral resistance to commercial building frames, the bents of columns and trusses themselves were called on to resist wind loads. Because of the ready availability of graphical methods, and their accuracy in determining forces in complex structures, these methods were frequently called on for the analysis of laterally loaded framed structures.

References Cited

Berlin Iron Bridge Company (ca. 1890). *The Berlin Iron Bridge Co., engineers, architects, and builders in iron and steel ...: office and works, East Berlin, Conn.* Case Lockwood and Brainard, Hartford, CT.

Bow, R. H. (1873). *Economics of construction in relation to framed structures*. E. and F. N. Spon, London.

Ketchum, M. (1903). *Design of steel mill buildings*. Engineering News Publishing, New York.

Shumway, A. (1904). *Wind stresses in a steel frame mill building*. BCE thesis, Cornell University, Ithaca, NY.

Sondericker, J. (1904). *Graphic statics, with applications to trusses, beams, and arches*. John Wiley and Sons, New York.

Part IV

Summary and Conclusions

16

Concluding Remarks—The Preservation of Historic Analytical Methods

In the introduction to this book, we identified three fundamental reasons for the importance of the study of historic methods of structural design. These reasons were, first, that understanding the intent of the designer is the key to a successful rehabilitation, whether architectural or structural. Second, the preservation of design methods for historic structures is at least as important as the preservation of the structures themselves. Third, many of the methods used in structural design in the late 1800s are valuable in their own right—quick, computationally efficient, understanding of the behavior of the structure, and often giving special insight into the actual performance of the structure. In this final chapter we will briefly review these reasons in light of the methods of analysis and design that have been introduced in the preceding chapters.

The analytical and design methods of the late nineteenth century are an embodiment of the spirit of the age and of the spirit of the engineering and construction professions. Although the growing application of scientific principles is observable in these methods, the application of science to engineering is inseparable from the use of empirical knowledge. To approach works from this period, it is necessary to have some understanding of the engineering methods used in their design and construction and particularly to understand the balance of analytical, graphical, and empirical analysis methods that gave rise to the structures that we now admire. For instance, it is hard to interpret the structure of a masonry arch bridge (Figure 16-1), or a Thacher (1884) truss bridge (Figure 16-2), without understanding the assumptions that went into their design and the conceptions of structural behavior that resulted in the selections of materials, member sizes, member

Figure 16-1. Wissahickon Creek Bridge, Philadelphia and Reading Railroad, 1881 (HAER PA,51-PHILA,698-).
Source: Photograph by Joseph Eliot.

Figure 16-2. Thacher truss bridge, Rockingham County, VA, 1898 (HAER VA, 83-BROAD,2-).
Source: Photograph by Jet Lowe.

configurations, connections, and supports that are reflected in the final design. The masonry bridge was designed by empirical or graphical methods, both of which have largely disappeared from the repertoire of the modern engineer—the bridge is not amenable to treatment as a framed structure, and the analysis of the arch is particularly difficult to undertake by modern methods. The Thacher truss, a hybrid of a Fink truss and a Pratt truss, is easiest to solve by the indexing method presented in Chapter 7 and challenging to solve by any modern method.

Instead of automatically using modern forms of analysis on a structure of these types, it is better to recognize and celebrate the widespread recourse of the profession to empirical and to semiempirical methods of analysis. We have noted that empirical design is a preferred method for a nineteenth-century designer: most decisions, even those arrived at rationally, were informed, enhanced, supplemented, and completed by empirical knowledge. Such knowledge often was used where analytical knowledge was lacking or where analytical methods provided misleading information. The empirical thinking of a nineteenth-century structural designer took three basic forms: the application of general ratios representing good design, application of other rules of thumb, and finally the simplification of complex analyses to the point of being manageable.

An example of the application of general ratios representing good design is the use of span/depth ratios as a means of establishing the size of structural members, either for preliminary or final design. This type of application is most often applied as preliminary design. Without necessarily quoting a span/depth ratio of approximately 15, an experienced carpenter recognizes that 2 × 8 wood floor joists can span 10 ft, 2 × 10 12 ft to 14 ft, and 2 × 12, up to 16 ft. Many structural engineers are familiar with the rule for steel beams of one-half inch of depth per foot of span. All authors on iron girders advanced proposals for the general depth/span ratios that need to be observed, with values ranging from 1:10 to 1:15, the larger being generally appropriate for bridge girders. From there, the width of flanges and the depth of flanges were also stated to be according to proportioning rules.

An engineer who calls for #4 reinforcing bars at 12 in. spacing without doing any further analysis is relying on experience, bolstered by the certainty that this form of reinforcement is effective in instances similar to the case in question; in other words, he or she is practicing empirical design. Other professional examples can be presented. Steel channels applied as stair stringers are rarely designed explicitly: they are simply chosen as C 10 × 15.3 or C 12 × 20.7, primarily on the functional or dimensional requirements for the stair. The evidence that built-up girder design was completed by sizing the flanges for the required resisting moment, neglecting the contribution of the web, is an example of the kind of simplifications that were practiced in nineteenth-century engineering.

Empirical design is also present in the imposition of appropriate minimum sizes, such as George Fillmore Swain's (1896) statement that the minimum web thickness of a built-up box member is 5/16 in. Although these minimum values were disputed or ignored by some manufacturers, such as the Berlin Iron Bridge Company, they are an instance of overruling the results of an analysis by the exercise of engineering judgment, formed on the basis of experience or skill.

The most persistent forms of empirical design involve some component of rational design presented as empirical formulas. Such a formula either involves the simplification of a rationally based procedure or curve-fitting to experimentally determined data. The first type of formula is a universal feature of building codes, both in the nineteenth century and in the twenty-first century. Ready examples of such a procedure are available in ASCE 7 (2010), in the determination of wind loads on buildings, or in the AASHTO *Standard Specifications for Highway Bridges* (2013), in the formula for the determination of distribution factors to girders, which was determined by multiple regression, not on experimentally determined data, but on data determined by finite element analysis in a parametric study. The nineteenth century furnishes similar examples, such as the rules by Robert Griffith Hatfield (1871) or John Daveport Crehore (1886). Similar rules are stated by Frank Kidder (1886) for the application of engineering principles by architects and builders. The other type of empirical formula has a basis in the collection of experimental data. In the nineteenth century Eaton Hodgkinson (1857) collected data from a limited number of tests and determined an empirically based formula for the strength of short and long columns. In the current century, our application of the Euler formula and the transition between elastic buckling and inelastic buckling reflect some interpretation of the results of testing, and the application of exponential laws to the determination of an interpolation curve between elastic and inelastic column behavior, which is based on probabilistic interpretations of empirical data (AISC 2011). The Rankine-Gordon formula in the nineteenth century can either be viewed as an analytical formula with the coefficients predetermined on the basis of the strength and stiffness of the material in question, or as an empirical formula with the coefficients of the formula to be determined empirically.

A critical look at empirical design invites comparison with contemporary engineering practice. Although twenty-first century engineering relies on a significant component of analytical thinking and analytical procedures, there remains a base of empirical design. Beyond its necessity for preliminary design, the use of empirical design persists throughout the design process. The application of proportioning is widespread, from Table 9.5 of ACI 318 (2011), through the evaluation of preliminary designs on the basis of span/depth ratios, to the customary proportions assigned to concrete and steel beam cross sections, to the adoption of slenderness ratios for columns that generally fall within predictable limits.

The analytical methods of the nineteenth century further test the assumptions of twenty-first century analysis. One example of a divergence between modern and nineteenth-century methods is the construction of column curves. In the present specifications for steel and wood, this is a cumbersome procedure, with the application of two formulas—a straight line yield ceiling and a hyperbolic Euler curve—with an empirical interpolation function between the two. The formula used in the nineteenth century represents a different viewpoint of the same problem. A single curve with most of the characteristics desired can be produced based on a different set of considerations. The contemporary application of the buckling limit state for columns leads to other incorrect conclusions. Although often analyzed as such, columns are never pinned at the ends: the flat surface on which they bear produces some rotational restraint, observable in tests of flat-ended columns. What is the reason for this predominance of the buckling limit state in the analysis of intermediate and long columns? Apparently, the understanding of the engineering profession of the early twentieth century suggests that this shift was motivated by a desire to use more rational and scientific methods. Euler buckling theory is significantly more advanced mathematically than the Rankine-Gordon theory of column strength, but the mathematical sophistication does not necessarily make it a better theory to apply to problems of columns buckling in actual structures, and, of course, the application of this rational theory requires the further use of an empirical formula in the transition between elastic and inelastic behavior.

The application of indexing methods to truss analysis, or the tracing of loads through the truss as practiced by Robert Henry Bow (1874), although sometimes less systematic, is a much more efficient method for analyzing a truss than the method of joints. Particularly in the case of a parallel chord truss, developing the indices for the forces in all the bars is a rapid process and promotes the visualization of the flow of forces through the truss. Similar tactics can be used for the analysis of pitched top chord trusses. However, for standardized trusses, such as the Fink truss with a 6:12 top chord pitch, the forces in the bars can be deduced almost instantly or calculated once in terms of panel length and load. By the end of the nineteenth century all of these tactics made the design of trusses very expedient. The analysis of trusses in the present day need not be any more difficult than it was 100 years ago. The charts developed for various truss types and printed in references in the nineteenth century (shown, for instance, in Figure 7-10) are equally applicable today.

The analysis of portal frames was done by approximate methods that eventually gave rise to the portal method. For a single portal frame, the essence of the portal method was simply to distribute half the lateral force to each column. This procedure only created problems when the windward column also was loaded laterally in addition to the roof truss or frame. The analyst then had to choose between Ketchum's method, in which each column's horizontal reaction is half the total lateral force, or Sondericker's (1904) method, in which the distribution of wind force to the two columns depends on equal deflections in the columns. This analysis, though, depended on a limited number of cases and could be readily reduced to a few formulas to be followed in different cases, such as base hinged or base fixed. The formulas that are available from Ketchum or Sondericker are still effective and may still be used for the design of a laterally loaded frame.

Graphical analysis was extraordinarily well developed by the end of the nineteenth century and fell into decline after the turn of the century. It was still applied to the design of trusses and was still taught in engineering schools through about 1950, but, as an analog method, graphical analysis of trusses was finally supplanted by the widespread use of the digital computer. The most compelling form of graphical analysis is surely the analysis of statically determinate trusses. In a single self-checking, self-correcting diagram, it is possible to infer the forces in all of the bars of the truss under a given loading condition. The diagram can be used to construct tables of bar forces under different conditions of load and panel length, and the modifications required by different slopes in the top chord can, in many cases, be introduced easily.

Less frequently used, but equally compelling, are the applications of graphical analysis to the design of beams. In these methods, the loads on the beams are used to construct a diagram that instantly results in the bending moment diagram for the beam and, based on Culmann's theorem, can be applied to the numerical determination of bending moments in the beam. Through the intervention of various other forms of analysis, it is possible to extend these methods to the analysis of beams continuous over several supports, although these methods were only used infrequently. Possibly the most often used of these methods was the method of Charles Ezra Greene (1877) for the determination of the bending moments in continuous girders.

Graphical analysis was particularly effective in the analysis and design of masonry arches. The ability to trace directly a (statically admissible) thrust line for any structure, including a masonry arch, was a defining feature of graphical analysis, and this method was widely employed and refined for the analysis of masonry arches. The contributions of Méry's

(1840) method to this analysis were significant and were often used as a means of reducing the number of trial thrust lines required to be drawn. The method was suitable for large bridge arches and for small arches framing over openings in the walls of buildings. Although analytical methods did exist for arches, they were cumbersome and depended on large tabular computations.

The genius of nineteenth-century engineering was the effective combination of these three fundamental methods—empirical, analytical, and graphical—and the selection of applications that were best suited to each method. Graphical analysis was universally adopted for trusses in buildings that had inclined top chords, less amenable to an analytical treatment, whereas analytical methods were almost universally used for bridge trusses, which tended to have parallel chords and for which the multiplicity of loading conditions made graphical analysis more cumbersome. Of course, empirical methods also were used for trusses, especially in the matter of appropriate span/depth ratios and sensible top chord pitch (the result of which is a span/depth ratio, e.g., 6:12 that is equivalent to a span/depth ratio of 4, which is about right for a pitched top chord metal truss) and were applied as an empirical principle to the design of trusses for mill buildings, the application of camber to the bottom chord of trusses, and especially to the development of connection details for trusses. Graphical methods also were used in trusses for the determination of deflections. Similarly, for beams, analytical or semiempirical formulas were sufficient for the design of most beams, but the determination of the maximum moment in a beam subjected to difficult loading patterns (such as produced by a locomotive) called for graphical analysis. Conversely, the production of ordinary wood or iron beams allowed the application of empirical methods, simply prescribing a span/depth ratio, and semiempirical methods, such as the collection of rules assembled by Hatfield (1871) or Kidder (1886). The design of ordinary arches was primarily empirical and the design of unusual arches primarily graphical.

Among the procedures of the nineteenth-century engineers that we have investigated, some appear to be potentially useful to twenty-first century engineers. Surely, the earlier engineers' facility with the design and analysis of trusses could prove useful to the contemporary engineering profession: not only in the type of truss to use for various building types and the methods to use to stiffen the overall building frame, but also in the effective analysis of the truss for gravity loads and the effective analysis of the truss/column/knee brace system under wind loads. The indexing methods for bridge trusses outlined in Chapter 7 can certainly be used as a means of checking building truss designs, and they have significant utility for the assessment of existing truss bridges.

Some of the other procedures used by nineteenth-century engineers also can be understood to be particularly efficient. The simplicity of the Rankine-Gordon formula for columns and its ability to encompass most column characteristics is worth noting. Moreover, the observation made by the earlier century's engineers that square-ended columns actually show some characteristics of fixed columns could be incorporated into column design to advantage, especially for wood columns. The National Design Specification (American Wood Council 2006) ignores this effect but supposes that a square ended column is actually pinned as if it were provided with a hinge.

Although arch analysis is rarely needed for new construction, the author is aware of several instances where a nineteenth-century method could be applied to advantage for the analysis of an arch for assessment of a bridge from the nineteenth century. Attempts to subject this type of bridge directly to a current method of analysis usually result in erroneous and overly conservative results. In keeping with the discussion on empirical design, it is first necessary to understand the construction characteristics of the arch, particularly on the use of haunching or filling, which is present in most arches of this period. Following this, an analytical or a graphical method can be particularly useful in determining the forces in the arch. Arches are particularly amenable to graphic analysis. The adoption of a method such as Méry's method has a long history in the application to the analysis of masonry arches, and its usefulness has not diminished. Although, as a practical matter, the statically indeterminate nature of the problem can be solved only qualitatively, the application of graphic methods can establish effective limits on the redundant quantities and can surely verify the safety of the arch as effectively as any other analytical method.

Graphic analysis, in general, is a neglected tool in the application of structural analysis to structural design. The ability to draw a shape for a structure appropriate to the loading that the structure is carrying could be particularly valued by engineering designers, and the resulting forms, while possibly regarded as innovative, will, in fact, represent the spirit of the engineers of the late 1800s who thought primarily through graphic structural analysis.

Finally, without being used directly as a design method, a greater recourse to empirical knowledge can be useful to any structural engineer. It is possible empirically to determine or check most design output, either by the use of span/depth ratios or by the simple application of practical experience. In either case, the review of how these methods were applied in the design of structures can have an effect on the way that a contemporary engineer works. Surely in the late nineteenth century this point was well understood.

References Cited

AASHTO (American Association of State Highway and Transportation Officials). (2013). *Standard specifications for highway bridges*. Washington, DC.

ACI (American Concrete Institute). (2011). "Building code requirements for structural concrete," *ACI 318*. ACI, Farmington Hills, MI.

AISC (American Institute of Steel Construction). (2011). *Manual of steel construction,* 14th Ed. AISC, Chicago.

American Wood Council. (2006). *ASD/LRFD, NDS, National design specification for wood construction: With commentary and supplement*. American Forest and Paper Association, Washington, DC.

ASCE. (2010). "Minimum design loads for buildings and other structures," *ASCE 7–10*. ASCE, Reston, VA.

Bow, R. H. (1874). *A treatise on bracing*. Van Nostrand, New York (originally published in Britain in 1851).

Crehore, J. D. (1886). *Mechanics of the girder*. John Wiley and Sons, New York.

Greene, C. E. (1877). *Graphical method for the analysis of bridge trusses*. John Wiley and Sons, New York.

Hatfield, R. G. (1871). *The American house-carpenter,* 7th Ed. John Wiley and Sons, New York.

Hodgkinson, E. (1857). In Tredgold, T. *Practical essay on the strength of cast iron and other metals*. J. Weale, London, 1860–1861.

Kidder, F. (1886). *The architects' and builders' pocket-book*, 3rd Ed. John Wiley and Sons, New York.

Méry, M. E. (1840). Sur l'équilibre des voûtes en berceau. *Annales des ponts et chausées*. 19, 15–70, and plates 133–134 (in French).

Sondericker, J. (1904). *Graphic statics, with applications to trusses, beams, and arches.* John Wiley and Sons, New York.

Swain, G. F. (1896). *Notes on the theory of structures,* 2nd Ed. Mimeographed lecture notes. Massachusetts Institute of Technology, Department of Civil Engineering, Cambridge, MA.

Thacher, E. (1884). *Bridge trusses.* Van Nostrand, New York.

Index

Page numbers followed by *b*, *f*, and *t* indicate boxes, figures, and tables.

About the Author

Thomas E. Boothby, Ph.D, P.E., R.A., is a professor of architectural engineering in the Department of Architectural Engineering at the Pennsylvania State University, where he has taught since 1992. Following the award of his B.A. and M.S. degrees in 1982, he practiced engineering and architecture in St. Louis, MO, and Albuquerque, NM, where he worked on various highway, building, and bridge projects and in building repair and rehabilitation. During this time, he earned registration as a professional engineer and an architect. He continued his studies at the University of Washington, earning a Ph.D. degree in civil engineering in 1991, and began his teaching career at Penn State. While at Penn State, he has taught structural engineering to both engineering and architecture students, and he has assisted state and county bridge engineers with the assessment and rehabilitation of masonry arch bridges. He has also completed research into the engineering aspects of medieval and nineteenth century buildings. He is currently researching the application and continuing usefulness of empirical design.